A color guide to familiar
MINERALS
AND ROCKS

By Dr Jiří Kouřimský
Illustrated by Ladislav Pros

OCTOPUS BOOKS

OCT 178

Translated by Zdenka Náglová
Graphic design: Soňa Valoušková

English version first published 1975 by
OCTOPUS BOOKS LIMITED
59 Grosvenor Street, London W 1

© 1974 Artia, Prague

ISBN 0 7064 0408 4

Distributed in Australia by
Rigby Limited
30 North Terrace, Kent Town
Adelaide, South Australia 5067

Printed in Czechoslovakia

3/02/28/51

HU

CONTENTS

FOREWORD

Collecting of all kinds is one of the most common hobbies of man all the world over; and in recent times, the collecting of minerals — although somewhat complicated because of its scientific nature — has been attracting more and more attention.

In addition to highly professional and scientific works on mineralogy, the literature in every language includes books written at a popular level and compiled solely for the interest of the common reader. The demand for such literature appears to be on the increase, as the number of those interested in the mineral kingdom is growing continually. We are therefore publishing this additional handbook on the subject. Its aim is to arouse initial interest and supply the reader with an elementary knowledge of this important branch of science.

The book is divided into two parts: the first contains an introduction to general mineralogy and a chapter dealing with the origin and occurrence of minerals throughout the world. The second includes pictures of the most important minerals from the main world and European sources. The book contains a small terminological vocabulary and a list of recommended further reading.

In the pictorial section, minerals are not arranged according to any system other than their importance in practical application. In the first place, mineral ores are cited (e.g. ores of iron, manganese, nickel, chromium, wolfram, molybdenum, tin, lead, zinc, copper, silver, mercury, gold, antimony, arsenic and uranium); these are followed by other, applied, minerals (raw materials applied in chemical industry, heat-resistant raw materials, raw materials applied in the ceramic, glass and building industries, including precious stones) and, finally, the most important rock-forming minerals.

Illustrations in colour show mineral specimens from the Mineralogical Section of the National Museum, Prague. Particular attention was paid to the selection of appropriate samples demonstrating the natural beauty of minerals and to the common samples which may be collected easily by everyone.

It is the author's wish to win admiration for the mineral kingdom, not only among collectors but also among all those who love the world of nature.

ELEMENTARY TERMS

The rocks which are commonly used as building stones can be seen on closer inspection to contain various minerals. Under magnification the rough surface of the rock reveals grains of differing colours, shapes and sizes, which are the individual minerals. Sand generally consists of colourless, transparent, more or less rounded grains of quartz. Soft scaly flakes of mica are also found glistening in sand. Both quartz and mica are minerals.

A *mineral* is a naturally occurring chemically and physically homogenous inorganic substance in a solid, or more rarely, liquid state, having a definite atomic structure. From the chemical viewpoint minerals are either compounds or elements. Naturally occurring substances are products which have arisen as a result of geological processes, unaided by man. Consequently, substances originating as a result of metallurgical processes, e.g. brass and steel, are not minerals. On the other hand, substances arising as secondary products on spoilbanks, e.g. slag heaps — which are the outcome of processes controlled by human agency — are considered to be minerals. Some mineralogists are even of the opinion that several organic substances, such as coal, amber and mineral oils, are also minerals.

The mineral is *homogenous* since the chemical as well as the physical composition of all its constituent parts is the same. In this way minerals differ from rocks which generally consist of a variety of minerals and form large parts of the Earth's crust. There are also rocks, such as limestone, calcite, marble and dolomite (which when pure are forms of limestone). These are called simple (monomineral) rocks. They are not considered minerals because they are not necessarily homogenous; they form large portions of the Earth's crust.

The science of the study of minerals is called *mineralogy*. The study of rocks is the subject of *petrography*, a branch of science related to mineralogy. *Palaeontology* is the study of impressions and remains of fossilised plants and animals sometimes displayed in rocks.

All these are branches of *geology*, the science of the origin, composition and development of the Earth, and are not only theoretical sciences — more and more they are affecting the daily life of modern man.

The relationship of man to the mineral kingdom of nature, his knowledge of the techniques of extraction, treatment and application of mineral materials has been — since the Stone Age — one of the leading factors affecting the evolution of human society. At the present time, minerals and rocks are fundamental industrial raw materials, whether they are ores yielding metals or raw materials applied in the metallurgical, chemical, glass, ceramic and other industries. More than 400 different minerals — out of a total number of about 1,800 more abundant minerals — are universally applied nowadays for industrial purposes. We cannot imagine modern life without the mineral raw materials.

Knowledge of mineralogy establishes the basis of some other sciences, such as *pedology*, the science and study of the properties of soils. Soils depend upon minerals comprising rocks through the decomposition of which they have been formed.

SHAPES OF MINERALS

Under favourable conditions, most minerals and many other solid substances assume certain definite geometrical forms — bounded by a number of surfaces — called *crystals*. Some crystals are so minute as to be almost microscopic; others may be of enormous size (several metres long). There are narrow crystals, acicular or needle-like, columnar, finely scaled or tabular ones.

Closer inspection of some crystallized minerals will show that all their crystals are almost identical in shape. Consequently, the shape of crystals is never accidental, but always governed by certain definite laws. The study and explanation of these factors, as well as the description of crystal forms of different substances, are the subject of *crystallography*.

Although, since Ancient Times, much attention had been paid to the study of crystals, especially their external form, crystallography did not become an independent branch of science until the 17th and 18th centuries.

The fundamental difference between crystals and amorphous substances is not just the external covering of the latter by crystal faces. In rocks crystalline mineral grains are irregular in shape and yet they are crystals. The basic difference lies in their internal atomic structure, i.e. in the arrangement of molecules, atoms and ions. In gaseous substances, liquids, and amorphous substances (such as glass), these particles occur in irregular, chaotic assemblages; whereas in solid crystalline substances they are arranged in almost parallel rows intersecting in a regular pattern. The crystal form is the outward expression of the internal structure of a given mineral. The physical properties by which a mineral is identified also depend on the internal structure.

Though some minerals crystallize in definite characteristic forms, the majority assume various shapes during the process of crystallization. For instance, calcite crystals occur in different,

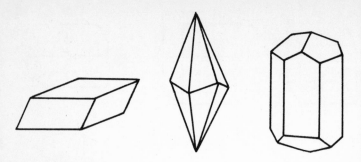

Fig. 1. Different forms of calcite crystals.

very often quite antithetical forms (high or low rhombohedra,
columnar, pointed or tabular crystals). All of them, however,
have a common feature, i.e. an identical symmetry corresponding to the internal symmetry of the groupings of the smallest
particles of the substance. According to this symmetry, crystals
may be classified into six large groups called *crystal systems* which
are named as follows:
(1) triclinic (some felspars)
(2) monoclinic (monoclinic felspars, mica, gypsum, augite,
    hornblende, wolframite)
(3) orthorhombic (aragonite, antimonite, marcasite, barytes,
    topaz)
(4) tetragonal (chalcopyrite, cassiterite)
(5) hexagonal (beryl, apatite) and rhombohedral (quartz,
    calcite, corundum, haematite, tourmaline, dolomite)
(6) cubic (fluorite, galena, rock salt, magnetite, garnet).
Crystals are divided within the individual systems according
to the axes of symmetry (see fig. 9.). These are a combination
of certain elements of symmetry; the highest symmetry exists in
the cubic system, the lowest in the triclinic system. In each
system crystals can occur only in the form characteristic of its
symmetry.

The main factors determining the shapes of minerals are their
chemical composition and the physical conditions of their
growth, especially temperature and pressure. During the process
of their growth, crystals do not always develop equally in all

12

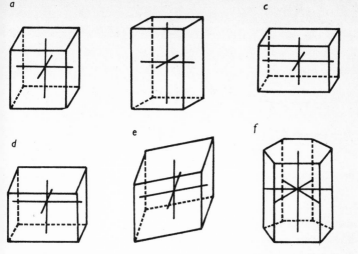

Fig. 2. Axial crosses of cube system (a), tetragonal (b), orthorhombic (c), monoclinic (d), triclinic (e), hexagonal or rhombohedral (f).

directions, the regular growth of crystal faces being restrained by various external agents. Thus, instead of being identical, crystal faces in one crystal become varyingly flattened or drawn out, or are even absent.

However, neither the size nor the shape of the faces are really essential for identification purposes. The essential factor is that the angle between individual crystal faces always remains the

Fig. 3. Hexagonal symmetrical crystals of water (snow flakes).

same. The measurement of these angles forms the basis of descriptive crystallography. Interfacial angles of crystals are measured by *goniometers*.

Some minerals display characteristic elevations and depressions on their crystal faces (accessories) arising in the course of the growth of crystals, either due to weathering or corrosion. They may therefore be of great importance in identification, e.g. the striated crystal faces of pyrite. The speed of growth of individual crystals differs considerably in different directions; so also do their physical properties (hardness, colour, optical properties, and electric conductivity).

If too many crystals are grouped together, they may interfere with one another's growth. If restrained in growth on one side, crystals grow on the remaining sides. This is why grains in rocks are of irregular shape. On the other hand, crystals growing isolated and alone in a soft and plastic environment may develop freely on all sides, with something of the regularity which the ideal model shows. Thus we find perfect crystals of gypsum in clays, of garnet and magnetite in various shales, or crystals in some igneous rocks (e.g. augites and amphiboles in felspars) which originated at the time when the solidifying rock resembled a plastic molten mass.

Crystals may also occur in cracks in solidified rocks as the product of precipitation from water, vapour or gases. Crystals adhering with one end to the rocky matrix substratum, possess well-developed faces only at the other end. Such conditions favour the origin of crystal groups.

Druses are groups of closely clustered, almost parallel, crystals growing from a common base. The parallel growth of crystals is possible only if the walls of cracks — the so-called druse cavities — are sufficiently even. If the cavity is more or less of globular

Fig. 4. Asymmetrical crystal of quartz.

shape so that it is lined with crystals pointing with their free ends to its centre, we speak of a *geode* (quartz, zeolites, calcite, etc.).

Minerals forming druses or geodes are a common feature. Minerals commonly found in druses are quartz, calcite, fluorite, barytes, etc. Excellent crystals of these minerals are often found in large cavities and cracks. In the so-called rock-crystal vaults in the Northern Urals crystals of 'rock crystal' were found weighing up to 500 kilogrammes. In summer 1945, a quartz crystal was found in the neighbourhood of Volhynia (Ukraine) weighing 10 tons. Similar crystals are also to be seen in the Alps; but the largest of all have been found in Madagascar.

In druses and geodes crystals are associated in more or less parallel groupings. Irregular clusters of crystals are called *aggregates*. They are very often composed of tiny crystals with only imperfectly developed faces. Such aggregates are called crystal aggregates. These may be differently designated according to the arrangement of their crystals. Thus, we distinguish granular aggregates (coarse-granular and fine-granular), columnar, acicular, hair-like or fibrous, tabular, lamellar, dendritic (arborescent), wiry, etc. Some aggregates of a more complex structure are designated by various special names. For instance, the *oolites* are masses of small rounded grains while *pisolites* are similar but larger. Calcite, haematite and other minerals may occur as pisolites or oolites. Such aggregates are composed of individual lamellae formed by microscopic crystals.

Apart from these aggregates, minerals occur in *compact masses* which are in fact parts of individual crystals exhibiting no external limits.

To describe tuberous (nodular) forms in sedimentary rocks the term *nodule (concretion)* has been applied in mineralogy and petrography. They are most commonly of a quartz composition such as chert, opal and chalcedony, and occur in limestones and serpentines. Also, some sorts of chalcedony and opal form nodules which are due to the deposition of different mineral substances accumulated about a centre of mineral grains, or fragments, of some organic matter. Sometimes, however, concretions have no core at all.

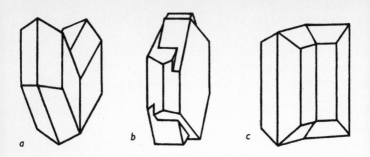

Fig. 5. Various crystal twins: a) gypsum, b) orthoclase (so-called Carlsbad Twin), c) aragonite.

A single crystal is sometimes made up of two or more individuals of an identical chemical composition arranged in accordance with some definite pattern. We distinguish the *parallel growth*, when several crystals have so intergrown that some of their faces are parallel, and *twins* (twinned crystals), when crystals intergrow at certain angles, and have some place or axes common to both. One half of the twin is produced by rotating the crystal through an angle of 180°. Intergrowths of three, four, five, or more individuals are termed threelings, fourlings, fivelings, and so on. According to the shape of the intergrowths, we distinguish *contact twins* (gypsum, aragonite), and *interpenetrated twins* (fluorite). Crystal twins are very common.

Much less frequent is the occurrence of the intergrowth of two different minerals of identical (pyrite and marcasite), or completely different, chemical composition (haematite-rutile).

A peculiar class of false forms are the *pseudomorphs*, which are quite common in nature. Thus we come across limonite in a perfectly bounded crystal form of pyrite. The crystal of pyrite was changed by oxidation and water absorption into limonit and retained its original crystal form. In a similar manner, a whole series of pseudomorphs originate, such as the pseudomorphs of malachite after pure copper or after azurite, limonite after siderite, etc.

Sometimes one mineral changes into another mineral of identical chemical composition. Thus, aragonite (orthorhombic calcium carbonate CaCO₃) changes into rhombohedral calcite, the crystal shape being retained. At times, one mineral may be deposited upon the crystal surface of another and enclose it completely. These are called *incrustation pseudomorphs*. In some cases the removal of the first mineral takes place simultaneously with the deposition of the incrusting substance. If the resulting cavity gets filled we speak of a *substitution pseudomorph*; it is conspicuous when the incrusting substance has been dissolved. In this way the pseudomorphs of quartz after calcite, or talc after quartz are formed.

In nature, much rarer than crystalline minerals are minerals which do not crystallize. They are called *amorphous* minerals. They are without definite form and internal structure and, in the course of time, they change into crystalline substances. They occur most often as incrustations or coatings, but may also originate in some other manner, such as limonite and opal — which originate by deposition from water.

If water saturated with iron and manganese oxides penetrates cracks in rocks, it leaves behind branching, fernlike structures. These phenomena are called *dendrites*, and are also amorphous minerals.

PROPERTIES OF MINERALS

The regular arrangement of atoms in the mineral determines its physical properties. It is most evident in the case of graphite and diamond. Both are composed of atoms of carbon, yet their internal structure is totally different. This shows that there is a certain definite relationship between the crystal shape and the physical properties of minerals. Minerals differ from each other by the shape of their crystals as well as by their properties, the latter being very important in their recognition.

The following chapter will deal with the main physical properties except those whose determination requires more thorough theoretical studies, or some special measuring apparatus. Here belong, for instance, various optical properties which are very important in a precise recognition of minerals. Let us mention at least one optical property, *double refraction*, which is typical of many crystallized minerals. If a cleavage-fragment of a very pure Iceland Spar (calcite) crystal in the shape of a perfect rhombohedron is placed over a handwritten page, it will be found that two images of the handwriting are seen. If the rhomb

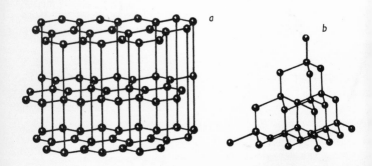

Fig. 6. Crystal structure of graphite (a) and diamond (b).

is rotated, one of these images remains stationary whilst the other moves round. This is due to a double refraction of the ray of light entering the calcite crystal. Double refraction is best detected by means of special microscopes, and may play an important role in the identification of even very small mineral grains.

There are, however, some optical properties which can usually be recognized upon first sight, such as transparency, colour and lustre.

Transparency

When a ray of light strikes upon a mineral surface a part of it passes through the mineral whereas the rest is usually reflected. The more rays pass through a mineral, the greater its transparency. According to the degree of transparency we distinguish:

Transparent minerals, which, irrespective of their thickness, transmit light easily (rock crystal, topaz, etc.). If they are colourless, they are indicated as pure.

Subtransparent or semitransparent minerals, when objects seen through them appear indistinct.

Translucent minerals, which are capable of transmitting light, but cannot be seen through (chalcedony, chert, flint).

Non-transparent minerals, which do not transmit light when in thicker slabs. When cut into very thin sections, however, they become transparent (amphibole, augite, schorl — the black opaque variety of tourmaline).

Opaque minerals, which absorb all rays of light so that no light is transmitted even in the thinnest sections (most sulphides, magnetite, etc.).

The general conclusion to be drawn is that the greater the density, the poorer the transparency, and vice versa.

Colour

Some minerals have a fairly constant colour (malachite, azurite etc.) whereas others occur in many different colours. According to the colour, we distinguish the following kinds of minerals

Idiochromatic minerals, which always occur in one and the same colour varying occasionally only in shade. The colour is then an inherent property of the mineral and is due to a certain definite chemical composition and a definite internal structure (haematite, limonite, magnetite, chromite, cinnabar).

Colourless minerals are minerals which in their chemically purest varieties are colourless (diamond, rock salt, rock crystal topaz).

Coloured minerals originate due to the pigmentation of pure minerals by different impurities. In minerals of a simple chemical composition — expressed by a definite invariable formula — colour is caused by mechanical distribution of some pigment At other times, the colouring may be due to the distribution of very small foreign mineral particles (interpositions) discernible only by means of a microscope (e.g. some feldspars).

A special kind of mineral colouring is the so-called *iridescence* It is exhibited by opaque minerals of metallic lustre, such as pyrite, galena, chalcopyrite. Moreover, these minerals exhibit the so-called *iridescent tarnish* on their surfaces; this tarnish is due to the incidence of light on the thin lamellae formed by chemical changes on the surface of the mineral (e.g. haematite, limonite).

The colour of minerals differs greatly when viewed by transmitted light from different directions (e.g. tourmaline). This property is called *pleochroism*. Some opaque and some translucent minerals display different colours in transmitted and reflected light. For instance, crystals of fluorite in transmitted light appear green to the eye whereas in reflected light they are blue violet.

In the identification of minerals it is very important to distinguish between the true colour of the mineral and its apparent colouring. The true colour of minerals is tested by means of the *streak*, which is the fine mineral powder produced by scratching the mineral. It may be provided by rubbing the mineral on the streak-plate, i.e. a piece of unglazed porcelain. It is often important to test the colour of the fine powder since it may be quite different from that of the mineral in mass. The colour of the streak, unlike that of the whole specimen, is constant and typical of the given mineral.

Lustre

The colour of a mineral is closely related to its lustre, which is due to the reflection of light from the mineral surface. Lustre depends upon various factors, especially upon the number of rays absorbed or reflected.

The kinds of lustre distinguished are as follows:

Metallic lustre, characteristic of the majority of opaque minerals, especially native metals and their sulphide ores. It is most perfect upon freshly fractured surfaces and crystal faces (silver, galena, pyrite, haematite).

Submetallic lustre occurs on well-crystallized dark, almost opaque minerals (wolframite, haematite).

Adamantine lustre occurs solely on transparent minerals with a high refractive index (diamond, sulphur). *Metallic-adamantine* is a term used to describe a variety of the adamantine lustre verging upon metallic (some sorts of sulphides, such as sphalerite).

Vitreous or glassy lustre is most frequent on transparent and subtransparent minerals. It is common in the crystallized forms of quartz, limestones and in most silicates.

Pearly lustre, or the *lustre of mother-of-pearl*, is shown by transparent or subtransparent minerals where the mineral has a good or perfect cleavage, such as some micas.

Resinous or waxy lustre is shown especially by opals.

Greasy lustre, similar to the resinous variety, is shown especially by minerals displaying numerous microscopic inclusions, such as quartz, apatite, uranite.

Silky lustre is peculiar to minerals having parallel fine-fibrous structures, as in varieties of gypsum (selenite and alabaster), fine-fibrous asbestos, especially the serpentine variety (chrysotile), fibrous limestones, etc. In comparison to other kinds of lustre it is less intense.

The above optical properties, especially colour and lustre, are sometimes designated as the *mineral modification*. Thus minerals are divided into those of *metallic modification* and those of *unmetallic modification*.

Density

Another physical property of minerals, used in their identification, is their density. Unlike the optical properties it is independent of the crystallographic direction. It is a ratio of the weight of the body to that of an equal volume of water.

The density — which is a good aid to identification — may be determined by various methods. An exact determination requires relatively expensive apparatus, such as a Jolly or Walker balance. Consequently, it will not be discussed in this book. Yet even casual collectors may succeed in approximate determinations by means of heavy liquids which show whether a mineral fragment floats or sinks in a liquid. The most important heavy liquids are the following: bromoform (2.88), tetrabrommetan (2.95), Thoulet's Liquid (3.19), methyleneiodide (3.3), Rohrbach's Solution (3.58) and Clerici's Solution (4.2).

Apart from the density — independent from the crystallographic direction — all crystallized substances possess *cohesion properties*, i.e. cleavage and hardness, which are closely related to the crystalline form.

Cleavage and Fracture

Cleavage is a characteristic property of crystals. It may best be observed on mica, which can easily be separated by a knife blade or nail into thin laminae or sheets. However, mica will only cleave in one direction, i.e. mica cleaves only along certain cleavage planes. The crystals of rock salt or galena, struck to

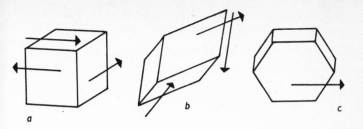

Fig. 7. Different cleavage of minerals: a) cubic cleavage of galena, b) rhombohedral cleavage of calcite, c) cleavage of micas in one direction.

whatever small fragments, always retain the shape of the cube. This means that they cleave only along straight, mutually perpendicular planes. Whereas glass, like other amorphous substances, breaks easily in all directions, into irregular pieces.

Cleavage is closely related to hardness; it is the ability of crystals to separate readily in certain definite directions under the effects of some mechanical force.

Crystals therefore cleave in directions of the least cohesion. Different minerals have different cleavage. Mica and gypsum possess a more perfect cleavage than rock salt, galena or limestone. Fluorite and amphibole have a difficult cleavage; an imperfect cleavage is obtained on garnets. A complete absence of cleavage is observed on quartz. For the determination of some minerals, not only the perfection of cleavage, but also the angle between individual cleavage planes (such as in augite and amphibole) are characteristic.

Bodies obtained by cleaving some minerals bounded on all sides by cleavage planes are called *cleavage forms*.

Under similar mechanical conditions amorphous substances exhibit uneven fracture surfaces generally indicated as *fracture*. According to the character of the surface, the following types of fracture may be distinguished:

Conchoidal fracture is the most common one. It exhibits curved surfaces, shell-like in character (quartz, opal, most silicates).

Uneven — the surface is rough by reason of minute elevations and depressions (pyrite).

23

Even — fracture surfaces are flat or nearly so (jasper).

Earthy — usually occurs in weathered substances, characteristic of soft and earthy minerals (kaolin).

Hardness

The resistance which the smooth surface of a mineral offers to a point or edge tending to scratch it is its hardness. The hardness of individual minerals is a constant quality. It is indicated relatively, in terms of Mohs's scale, arranged by the German mineralogist Friedrich Mohs (1773—1839). The scale consists of ten common minerals arranged in order of increasing hardness as follows: (1) talc, (2) rock salt (or gypsum), (3) calcite, (4) fluorite, (5) apatite, (6) feldspar, (7) quartz, (8) topaz, (9) corundum and (10) diamond. The grades 1—10 serve for comparison purposes; the values assigned to the members of the scale indicate simply the relative hardnesses.

When using this scale we attempt to make a scratch on each of the scale minerals in turn with the specimen being tested. Determination of the approximate hardness is possible — irrespective of the scale — by applying the following technique: minerals corresponding to (1) in hardness can be scratched by fingers (graphite), (2) can be scratched by a finger-nail (antimonite), (3) by a copper coin (muscovite), (4) and (5) by a knife (dolomite) — minerals of (5) correspond to the hardness of glass, e.g. opal, (6) and (7) can be scratched by a steel file (chalcedony). Minerals of hardness (6) can scratch glass, those harder than (7) emit sparks when struck with steel. Minerals having a hardness more than (8), i.e. 8—10, are comparatively scarce.

For an exact determination of hardness a *sclerometer* is used. It is provided with a diamond or steel point for scratching. Hardness is determined by weighing the point with weights.

The exact determination of hardness is impossible on earthy, fibrous or laminated aggregates because of their porosity, which makes them seem to be softer.

Other Physical Properties

For minerals occurring in the form of thin laminae (sheets), *elasticity* or *flexibility* is a very important property. Flexible minerals are those that bend easily and stay bent after the pressure is removed (e.g. chlorites), in contrast to elastic minerals (e.g. mica, which in other instances strongly resembles chlorites), which when bent spring back to their former position.

Another property is *solubility*. If a crystal dissolves only partly, regular etching shows on its faces. The etching of minerals is often carried out in order to find out the actual symmetry of the crystal which is not always apparent from its external shape.

A very interesting character of some minerals is their *radioactivity*, an invisible emanation which is due to the natural radioactive decay of some elements. It was noticed for the first time in 1896 by the French physicist H. Becquerel on uranium from Joachimsthal (now Jáchymov in Bohemia). The radioactivity is determined and measured by a special measuring unit, the Geiger-Müller counter. Radioactive minerals, e.g. uraninite, contain radioactive elements, especially uranium and thorium; some other minerals contain these elements in traces (e.g. some zircons).

Another interesting feature is *magnetism*. There are only a few actively magnetic minerals, i.e. minerals which themselves attract other metallic objects (e.g. magnetite). On the other hand, there are minerals containing large amounts of iron, such as haematite, which are attracted by a magnet.

In conclusion we should also mention *luminescence*. It is a property enabling minerals to achieve conspicuous shining colours when irradiated by ultraviolet or X-rays. For instance, calcite, some kinds of corundum, orthoclase, tourmaline, apatite, fluorite, scheelite, etc., may display intensive luminescent colours. This colour is not always the same in one and the same mineral, but differs, sometimes greatly, according to the admixtures contained in the mineral. There is a difference between the luminescence of the blue corundum — sapphire — and its red variety — ruby. Very often, however, different luminescence is exhibited by sapphires and rubies from different sources. This property may successfully be applied in the study of cut precious stones, in determining their place of origin, or in distinguishing between real and synthetic precious stones.

ORIGINS AND OCCURRENCE OF MINERALS

Minerals originate, undergo conversion and decay just as do the living parts of nature. Mineralogy studies and explains all these changes. To understand the natural laws governing the conversion of minerals in nature, these processes must not be studied and judged separately. The whole mineral family, i.e. *paragenesis*, must be considered.

Similarly, as with various chemical elements, so various minerals are represented in the Earth's crust in different quantities. Many minerals may be found almost universally. On the other hand, some occur in nature very rarely, existing only in certain mineral families (paragenesis), e.g. certain silicates are formed solely from magma and are consequently found especially in igneous rocks.

The occurrence of minerals in nature is regulated by the geological conditions throughout the life of the mineral. A particular mineral may only be found in association with one rock type, e.g. granite, or may be found associated with both igneous and sedimentary rocks. Similar physico-chemical conditions may be repeated throughout geological time so that the mineral associations are also repeated.

Rocks as sources of minerals may be divided into three large categories:

Igneous rocks, formed by cooling of a liquid magma.

Sedimentary rocks, produced by the activity of water, wind, glaciers or gravity.

Metamorphic rocks, produced from igneous or sedimentary rocks by the action of high temperatures and pressure.

Igneous Rocks

At some depth within the Earth, molten material or molten rock substance occurs which is generally called *magma*. After cooling and solidifying, the magma forms igneous rocks composed of different minerals. How is it possible that the more or less

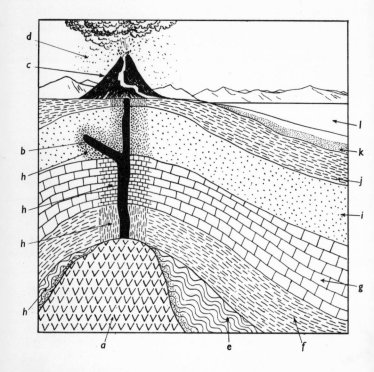

Fig. 8. Origin of different types of rocks:

Igneous rocks: (a) intrusive (plutonic) rocks consolidated at great depths, (b) vein rocks, (c) extrusive rocks (volcanic), (d) pyroclastic extrusive material blown from volcanic orifices.

Metamorphic rocks: (e), (f), (g) rocks changed by high pressures and temperature at different depths, (h) rocks altered by contact metamorphism in place of contact with molten eruptive material.

Sedimentary rocks: (i), (j), (k), (l) deposits dating from different geological ages.

homogenous magma, when consolidated, is composed of different minerals? Every chemical substance solidifies or melts at a certain temperature. In a similar manner every mineral component of the magma solidifies and crystallizes only at a certain temperature, e.g. quartz at 1713 °C, sulphur at 115 °C, orthoclase at 1300—1350 °C. If the magma within the interior or on the surface of the Earth becomes gradually cooled, chemical substances composing it do not solidify at one time but gradually do so according to the temperature at which they normally crystallize. The origin of different minerals is consequently governed by certain definite physico-chemical laws. In igneous rocks especially, minerals formed at high temperatures are to be found.

Intrusive Igneous Rocks

The most frequent intrusive (deep-seated) or plutonic igneous rocks are *granites*. They are composed of three main mineral constituents, i.e. feldspar, quartz and mica, which most often occur in the form of fine grains or plates, with only occasional larger crystals being found, in a fine-grained matrix. The formation of these crystals, the *phenocrysts*, under conditions of slow cooling was followed by more rapid cooling during which the fine-grained groundmass consolidated. Such texture is called *porphyritic texture*.

The commonest type of mica — occurring in plutonic rocks — is the white mica or muscovite, less frequent is the black mica or biotite, the latter occurring more often in the so-called basic plutonic rocks. (Basic rocks are rocks low in silica. Silica may occur either as pure SiO_2, i.e. quartz, or as the building component of the silicates.)

Granite is the commonest type of rock on the Earth. It is especially concentrated below its surface. Granites and the associated rocks occur in large quantities almost everywhere. They are cut in large blocks in granite quarries and are often used as building material. The largest granite mass measuring 23.000 km² is in Finland.

Another type of intrusive igneous rock formed predominantly of feldspar, quartz and white mica is *pegmatite*. They are coarse-grained quartz-feldspar rocks. In some of them mica may be absent, or white mica may be substituted by the black variety, or they may contain both kinds of mica. Apart from these main constituents, the pegmatites may contain some accessory mineral particles formed from vapours escaping after volcanic eruptions. The vapours contained some valuable elements, e.g. boron, fluorine, lithium, beryllium, or elements of the so-called rare earths (cerium, lanthanum, yttrium, etc.). Therefore desirable minerals, such as precious stones (in workable quantities), e.g. beryl, topaz, coloured tourmalines, garnets, or minerals containing precious metals, e.g. wolfram, molybdenum and uranium ores, or lithium potassium mica (lepidolite), or lithium biotite (zinnwaldite) may be found in pegmatites.

A special kind of pegmatite is *graphic granite*. Its two main constituents, feldspar and quartz, are intergrown so that their flat surfaces display a structure resembling cuneiform or Hebrew writing.

The most important areas where pegmatites occur are Brazil, the USA (North Carolina and Maine), India, Madagascar, and the Urals. The main industrial minerals of pegmatites are feldspars, quartz and mica.

Extrusive Igneous Rocks

Apart from intrusive igneous rocks — composed solely of crystalline minerals — there are magmatic rocks formed predominantly of consolidated non-crystalline mass, the so-called *volcanic glass*. They are volcanic ejecta, or effusive (volcanic) rocks, differing from intrusive rocks in structure as well as external form. They usually occur in volcanic areas where they have consolidated below the surface, or on the surface of the Earth.

The Alpine-Himalaya system formed by folding during the Tertiary contains good examples of volcanic rock. The mountains include the Pyrenees, Alps, Carpathians, Himalayas as well as the Atlas (North Africa) and Cordilleras (USA).

There are many kinds of extrusive igneous rocks having volcanic glass as a component. Moreover, there are also others — although less frequent — composed solely of this glass resembling the dark sort of artificial glass. These are called *obsidian*. They are black or blackish-brown, with conchoidal fracture. Very often they occur in the form of aphrolitic lava on the surface of consolidated lava flows. They are commonly found in active volcanic regions, e.g. in Italy, Iceland, Mexico, Indonesia.

Extrusive rocks may also occur in areas which originated in other periods than the Tertiary. The oldest extrusive rocks are the greenish spilites and diabases, which are of almost identical chemical and petrographic composition. *Spilites* are predominantly of the pre-Palaeozoic (Algonkian) Age, *diabases* of the Palaeozoic Age. Mineralogically, spilites are poorer than diabases, whose cracks and amygdaloidal cavities have been filled with some secondary minerals, especially calcite (in cracks) or iron ores, such as haematite and magnetite (in some places). In the late Palaeozoic (Permo-carboniferous) up to the Mesozoic Era eruptions of *melaphyres* took place. They are granular, dark grey-green, blackish-grey or red-brown rocks, occuring in Brazil, West Germany (Idar) and Bohemia (Giant Mountains). In their red-brown variety hot gases escaping from the magma formed comparatively large cavities filled later with different secondary mineral matter, such as various kinds of quartz and chalcedony.

The mineralogically richest extrusive rocks are the greyish *phonolites* and dark, almost black *basalts* — both mostly of the Tertiary Age. Their constituents are black crystals of augite and amphibole, biotite, olivine, less often feldspars or magnetites, as well as some other minerals. By cementation of fragmental basaltic rocks, basalt tuffs have been formed. Due to weathering, crystals of some minerals become loosened and may be found in many different places, including the arable soil, e.g. Kaiserstuhl (West Germany) and many places in the Central Bohemian Highlands, which are the source of augites and amphiboles.

Volcanic activity (e.g. in the Tertiary) gave rise to ore veins and ore deposits.* Prior to it the pre-existing rocks had been altered by the passage of hot (hydrothermal) solution and gases escaping from the Earth's interior. This process is called *propylitization*. This alteration was followed by the actual mineralization, i.e. the deposition of ore minerals from solutions. Many important ore deposits are therefore associated with the occurrence of extrusive rocks of Tertiary origin; many others occur in intrusive rocks. Ores occur most often in the form of veins or lodes, less frequently as lenticular bodies or as irregular masses. In many ore deposits *metasomatism* has played an important part, being the complete or partial replacement of a pre-existing rock by the ore-body. Metasomatic deposits may be found, for instance, in carbonated rocks which are dissolved by heated solutions and take over their ore content.

The most common ore occurrence are *ore veins*, which are of great economic importance. They have been formed in rock fissures, principally as a result of the solidification of mineral matter from solutions. Ore veins are divided according to their mineral composition into four main types: cassiterite veins, gold bearing veins, sulphide veins composed predominantly of sulphides, and iron and manganese ore veins.

Cassiterite veins usually occur cutting granites; their mineral composition differs only slightly from that of the granites and pegmatites. They contain cassiterite, some kinds of mica including uranium micas, and fluorite. Carbonates — which are quite common in other types of ore veins — are almost absent. Sulphides, such as molybdenite, sphalerite, chalcopyrite and arsenopyrite, are rather scarce.

---

* In the following sections we shall refer to ores and mineral raw materials. These terms are not exactly mineralogical or petrographic terms, yet under certain definite conditions they are applied to some minerals or rocks. Mineral raw materials are mineral substances which are applicable for technical purposes. Ores are mineral substances comprising metals in quantities of economic value which repay working. Consequently, the terms are relative. Precious stones may also be included in mineral raw materials (Author's note).

Gold-bearing veins are mineralogically poorer than the cassiterite veins. They are predominantly filled with quartz and pyrite, less often with antimonite or arsenopyrite. All these minerals contain gold only in finely disseminated grains. Its recovery still pays at 1—10 gr/ton. Gold-bearing veins are found in the USA (Mother Lode, California), in Australia (Mount Morgan), in the USSR, and in France. These large deposits are associated with the occurrence of granitic intrusive rocks.

Sulphide veins, that is veins composed predominantly of sulphides, are the most common. According to the metals contained in them they are further divided into veins and deposits of copper, silver, lead and zinc ores, antimony ores, and mercury ores. The occurrence of silver, lead and zinc ores is closely associated with the hydrothermal occurrence of uranium ores.

Copper ores occur either in veins or in bedded or lenticular deposits. Their principal source is chalcopyrite, which usually occurs associated with pyrite and arsenopyrite. In the locations in which they occur, apart from copper deposits, there are also some more valuable minerals, and minerals formed secondarily by the decomposition of the former. There are copper carbonates (malachite, azurite), copper silicates, and, less often, copper arsenates. Native copper usually occurs only sporadically in the upper layers of both previously-mentioned types of deposits. Copper ore veins are found, for instance, in Butte, Montana (USA), in the Lake Superior region, Michigan, and in Yugoslavia; bedded deposits occur in Zaire (formerly the Congo) and the USSR. The deposits in Zambia are similar to the Zaire deposits — which may be ranked with the world's richest — and are of a very complex and, so far, obscure origin. The most important deposit of copper ores is in Tsumeb, South-West Africa.

Silver, lead and zinc ores most often occur in association, in the same places, and are consequently called *poly-metallic ores*. Let us mention at least some of the world-famous deposits of these kinds of ores.

Kongsberg, Norway, has furnished a great deal of native silver from its large masses; similarly the Saxon mines at Freiberg — the centre of the ore-mining area of the eastern Ore Mountains in Saxony — were for a long time heavy producers. Th most important minerals are argentite, arsenopyrite, marcasi and fluorite, apart from a whole series of valuable silver ore Another locality is Schneeberg, famous for its deposits argentite and nickeline. On the Bohemian side of the O Mountains, in the famous mining town Jáchymov, the commc polymetallic ores are found. Moreover, there are also small quantities of nickel, cobalt, bismuth, arsenic and uranium ore The ore veins in the vicinity of Příbram, Bohemia, differ fro those of Jáchymov in possessing a higher content of antimor and its compounds, and in having a considerably lower conte of arsenic, uranium and cobalt-nickel ores. Another famo locality in this region is Kutná Hora, and the area of the Gia Mountains, where arsenopyrite occurs. All these minerals occ in association with granite rocks.

Ore veins occurring in Tertiary extrusive rocks may be fou in Europe, for instance, in the Carpathians. In Slovak especially in the area of Banská Štiavnica, sulphides are presented by large quantities of sphalerite, galena and pyri and, in places, marcasite and chalcopyrite; the occurrence argentite is also characteristic. Similar mines may be found the Carpathians in northern Romania, and in Transylvan which also yield polymetallic ores, especially lead ores. F instance, the deposits in the Siegen region (West Germany) a in the Slovak Ore Mountains are of metasomatic origin. O side Europe, polymetallic ore veins are abundant in Mexico, t USA, and Australia. A characteristic feature is their hi content of arsenopyrite.

Special attention should be paid to hydrothermal depos with *uraninite* as the main ore mineral. They may also be includ in polymetallic deposits. They belong to the most importa world sources of uranium, and are consequently of an extrem high value. They are found especially in Canada, Katar (Zaire), Portugal, England and near Tuya Muyun, Fergana, the USSR. In Shinkolobwe (Kasolo), Katanga — a typi source of uranium ores — uraninite occurs associated w cobalt-nickel and silver ores. Moreover, gold and other r minerals, such as the compounds of selenium, are found. T domestic secondary minerals formed by the decomposition of uraninite are world famous. They are of shining colours, especially the so-called uranium micas and uranium ochres (hydrated uranium arsenates, phosphates and sulphates).

Antimony ores, besides antimony, contain few other minerals. Antimony occurs predominantly in vein deposits, and only rarely in metasomatic deposits. It sometimes contains a considerably large amount of gold, less often lead, zinc, copper and arsenic. Large antimony deposits are in the Chayanta, Palca and Tupiza regions, Bolivia, San Luis Potosí, Sonora and Oaxaca states, Mexico, and in the Hunan and Yunnan provinces, China. The Chinese deposits are the largest deposits in the world, having 4/5 of the world production. The largest deposits of antimony in Europe are the Italian deposits in Sardinia and Tuscany, and the Yugoslav deposits in the vicinity of Fojnica, Krupanj and Kostajnik.

Mercury ores are predominantly composed of cinnabar, which usually occurs associated with quartz, carbonates, pyrite, marcasite, antimonite, etc. The most important producers of mercury ores in Europe are Spain, with the Almaden deposit (where cinnabar had been worked by the Ancient Greeks as early as in 7th century B.C.), Italy (Monte Amiata deposit), and Yugoslavia (Idria in Slovenia and Avala near Belgrade). The largest world producers are the USA, especially the states of Texas, Oregon and California, and Mexico together with the Chinese provinces Hunan and Szechwan. There are also important deposits in the USSR, especially in the Caucasus, Central Asia (Khaydarken and Chauvay), and in the Donets basin near Nikitovka.

Iron and Manganese Ores. Veins are predominantly composed of fibrous haematite and a mixture of manganese ores, i.e. manganite, pyrolusite and psilomelane. There are notable deposits of siderites penetrated by younger copper, bismuth, cobalt, nickel and lead ores in Siegerland (Rhine area). Similar deposits also occur in the Spiš-Gemerian Ore Mountains, Slovakia, and in Erzberg, Styria. Haematite veins occur in Egremont, England, and in the neighbourhood of Horní Blatná near Jáchymov, Bohemia.

Sedimentary Rocks

Many minerals originate as a result of deposition. Rivers carry the load of fine material, sand or larger particles. The greater the velocity of the stream, the larger the volume of eroded and transported particles. According to the velocity of the flow, the particles are deposited, that is, larger particles in the upper course, and smaller material and finer particles in the central and lower course of the stream. In the same manner, minerals are laid down on the sea floor due to the activity of glaciers, wind, and the accumulation of material of organic origin. The resulting rocks are called *sedimentary rocks*.

The most common sedimentary rock is *limestone*. Limestone rocks originated in almost all the geological periods, and are found almost everywhere in the Earth's crust, especially in the form of marine deposits. They are often composed of fragments or complete shells of various organisms, such as molluscs, sea urchins and corals. Limestone is therefore a typical organogenous sedimentary rock, that is a rock formed by the action of organisms or by their deposition. As a rule, limestone rocks form massive beds and long mountain ranges. They are known under several names. Loose limestones mixed with argillaceous or arenaceous material are called *marls*, solid varieties are *arenaceous marls* and *marlites*. *Chalk* is a soft white to light-grey earthy carbonate of lime deposited in the Late Mesozoic (Cretaceous formation).

Travertines or *Calcareous (Culc) Tufa* are Quaternary cellular deposits of calcium carbonate derived from fresh waters by the action of water plants, especially the algae, whose fragments they often contain. They are often used as building and ornamental stone and for decorative purposes, especially for decorating home interiors.

In fissures and cracks in limestone beds crystals as well as dripstones of calcite and other minerals may be found. They very often contain siliceous nodules composed of flint, chalcedony or opal-cacholong, formed of the hard parts of bodies of sponges living especially in Mesozoic seas. Rarer in occurrence are the crystallized kinds of quartz, such as rock crystal and amethyst. The best known places for flint nodules, associated with the pyrite or marcasite concretions, are the chalk cliffs on the Rügen island.

Under favourable conditions pure limestones may form the so-called *karst phenomena* (karst forms), which are due to the dissolution and leaching of limestones by water. In this way, extensive subterranean cavities, galleries, chimneys and caves have been formed, often displaying an attractive dripstone decoration. Well-known karst areas occur in Yugoslavia (Istria, Dalmatia), Greece, Turkey, in the Pyrenees, in Czechoslovakia, Belgium, in the USA (Mammoth Caves, Kentucky), Mexico, Jamaica, Java, China and the Caucasus.

Limestone beds occur associated with deposits of *bauxites*, that is, sedimentary rocks mainly composed of hydrated aluminium oxide. Bauxite deposits result from the decay and weathering of aluminium-bearing rocks under tropical or sub-tropical conditions. Bauxite may form residual deposits replacing the original rock, or it may be transported from its place of origin and form deposits elsewhere. The largest bauxite deposits occur in Central America, in Europe, France, Hungary, Yugoslavia and Greece.

Due to deposition from water, *sedimentary iron ores* have been laid down on the floors of seas and lakes. They are of different geological ages. There are many types of such deposits, some of them being of considerable importance. They are composed of haematite and silicates of iron. Equally important are the deposits of *manganese ores*, formed either by the manganese carbonates or manganese oxides. The latter are economically the most important deposits. They occur in Italy, in the USSR near Nikopol, and in the Caucasus.

Among the geologically oldest deposits, the so-called *alum shales* may be included. They are rich in pyrites and as late as the last century they were still extracted in many countries for the production of sulphuric acid. They are of Eozoic (Algonkian) or Early Palaeozoic Age. These rocks are interesting from the mineralogical point of view. Very often their sulphides are decomposed and give rise to different secondary sulphates.

Also *coal-bearing strata* are mineralogically interesting, especially the carboniferous and permian beds (late Palaeozoic). Coal originated as a result of the transformation (dry distillation) of pre-existing swamp plants, especially ferns. Coal layers often contain large quantities of pyrite and marcasite, or different concretions, such as pelosiderite, and other minerals. Thick deposits of brown coal originated also from plants in the Tertiary. They contain marcasite and pyrite, by the decomposition of which secondary sulphates, most frequently gypsum and melanterite (Green Vitriol), are formed.

Crude petroleum is approximately of the same age; most probably of organogenous origin. The most important deposits of petroleum are concentrated in two large areas: the American area, with deposits in the USA, Venezuela and Colombia, and the Near East area, extending to Siberia and the adjoining parts of Europe along the Caspian Sea.

The majority of minerals most probably originates by *deposition from sea water*, which contains many different salts, especially sodium chloride. In places with hot, dry climates, water in gulfs and lagoons evaporates, and crystallized salt is deposited on the sea floor — usually in basins with small freshwater feeders. In this manner, for instance, beds of gypsum and rock salt have originated. Classic examples of continuous rock salt deposition are the Salt Lake, Utah; the Dead Sea; Elton near Astrakhan and Kara-Bogaz-Gol, a gulf of the Caspian Sea.

The youngest sediments (Quaternary Age) are predominantly composed of soils, sands, gravels, loess (sediments deposited by wind), and loess soils. Rich sources of minerals occur especially in alluvial sediments where heavy insoluble minerals (gold, diamonds, garnets, sapphires) are concentrated. World-famous deposits are, for example, in India, Brazil and Zaire.

Minerals originating at the present time are called *recent minerals*. At the present time the youngest sort of coal — peat — is still being formed in many places due to the carbonization of peat-moss. Examples of recent minerals are the deposits at some hot springs, e.g. aragonite, opal-geyserite, and the so-called *guano minerals*, that is, phosphate deposits resulting from the accumulation of organic remains (bird- and bat-droppings). They occur in many karst caves.

Metamorphic Rocks

Rocks which originated at a certain depth below the surface of the Earth, or on the surface, do not remain unaltered. In the course of many tens of millions of years some strata have subsided or have been repeatedly flooded by the seas whose floors have gradually been covered with thick deposits. Thus rocks originally formed on the surface of the Earth were buried to great depths. They were then subjected to high pressures and high temperatures and altered to the so-called metamorphic rocks. In these rocks — as in igneous rocks — minerals formed under high temperatures are especially found.

In nature many areas are composed both of metamorphic and intrusive igneous rocks. They are then called *crystalline rocks* and the area of their occurrence is called the *crystalline basement*. In his system, metamorphic rocks occur associated especially with granites, pegmatites and other related rocks. Mutual transitions of both these types are quite frequent.

From metamorphic rocks poor in silicon dioxide the most important transitions are serpentines and amphiboles. *Serpentines* originated predominantly from the alteration of rocks rich in olivine. The best sources of serpentines are in the USA, Cuba, New Caledonia, and in Europe in Greece, Italy, Poland, Czechoslovakia, Saxony, Cornwall (England), and in the Urals. The occurrence of serpentines is important not only from the mineralogical point of view but also from the economic viewpoint. There occur in massive serpentines numerous small veins of chrysotile — serpentine asbestos, worked for its excellent quality as the best sort of asbestos. A further weathering of serpentines results in the origin of different hydrated silicates, e.g. chlorides, talc, magnesium carbonate — magnesite — and the hydrated silica — opal.

Another important group of metamorphic rocks poor in silica are *amphibolites*. They originated as a result of recrystallization of igneous rocks also with a low silica content, and are mainly composed of amphibole and soda-lime feldspars (plagioclase). Mineralogically interesting phenomena are fissures in amphibolites whose walls are decorated with druses of different secondary minerals containing especially sodium and calcium

Crystalline limestones or marbles, originally sedimentary rocks, were altered at great depths under high temperatures into crystalline rocks. They differ from the sedimentary massive limestones in their grain size. In many cases the recrystallization was accompanied by changes in chemical composition. It occurred especially along the contact line between the limestones and the igneous mass. On its way from the interior of the Earth, the magma often penetrates and invades the surrounding rocks, partly melting them, the massive rock being affected especially on contact with the magma. From the slowly cooling magma, gases and vapours escape through fissures and cracks in the Earth's crust. They are accompanied by solutions of different compounds which gradually dissolve the original rock and replace it with new minerals. At the same time, the original rock (most often the limestone) changes. Minerals originating along the contact line are termed *contact minerals*.

Closely related to crystalline limestones are *graphite deposits* often forming intercalations in crystalline limestones or in gneisses. Graphite deposits contain the same minerals as are found in crystalline limestones, only they are of a less typical form. Large quantities of pyrites are frequent.

Gneiss, the most common metamorphic rock, is mineralogically poor. Most often it contains different kinds of garnets or various aluminium silicates (the latter being characteristic of gneiss of sedimentary origin, the so-called paragneiss). In its composition gneiss resembles granite, its constituent parts — as in granite — being quartz, mica and feldspar. It can be recognized by a typical stratification.

The metasomatic deposits of carbonaceous rocks (metasomatism, see page 185) belong here by origin. They have resulted from the dissolution of the original beds of calcium carbonate (limestone) or carbonate of calcium and magnesium (dolomite), and their subsequent replacement by less soluble components. The most common mineral in such deposits and, at the same time, an important iron ore, is siderite. It is altered to limonite on the surface of the deposits. Limonite forms the major part of the so-called *iron gossan*, that is the zone of surface weathering of ores. In a similar manner also, the magnesite deposits — which are of considerable economic importance — have been formed. Both types of deposits occur, for instance, in Austria, Czechoslovakia and the USSR.

Of later origin, most probably, are the different sulphides penetrating the carbonate, and partly replacing them. Copper ores, e.g. chalcopyrite, are comparatively frequent, but also lead ores, pyrite, cobalt-nickel ores, cinnabar and antimonite occur. All these ores are worked, being of economic importance. As a result of their decomposition, a whole series of sulphates, carbonates, phosphates and arsenates are formed. They are treated in more detail in the chapter discussing the main ore deposits since they are closely associated with igneous rocks in origin.

Another type of ore occurrence, especially in ferruginous metamorphic rocks, are *skarns*, originating most often as a result of the alteration of limestones contained in crystalline schists. In these alterations, under the effect of neighbouring rocks, the limestones are gradually replaced by the so-called *skarn minerals*, e.g. magnetite, usually accompanied by iron-rich silicates (amphiboles, calcium iron silicate — andradite [common garnet], and other basic minerals). Less frequent are haematite and sulphur ores. The largest deposits of skarns are in Finland (Outokumpu), in Central Asia and elsewhere.

CHEMICAL COMPOSITION OF MINERALS AND MINERAL SYSTEMS

Every mineral has its own invariable chemical composition expressed by a chemical formula. In nature, however, almost no mineral occurs in its pure form but contains different admixtures. In many minerals these admixtures are not accidental since the occurrence of certain pairs or groups of elements is governed by strict physico-chemical laws. For instance, galena (PbS) usually contains silver as an admixture. The silver may be replaced by lead because of a close similarity in the size of the smallest particles in both elements.

Such replacement of elements in chemical compounds is called *isomorphous replacement*. In some minerals two or more elements are often replaced in different proportions. In olivines iron may be replaced by magnesium, and the other way round; their proportion (molecular weights) in various olivines is different. Therefore the formula for olivine — the magnesium iron orthosilicate — is as follows: $(Mg,Fe)_2SiO_4$, where the elements Mg and Fe, divided by a comma and given in brackets, can replace each other. Formulae of this kind are not common in chemistry. In mineralogy, on the other hand, they are necessary, since only in this manner may more complex relationships between individual elements be expressed. Many other minerals may be expressed by common formulae of chemical compounds (e.g. SiO_2 — quartz, $CaCO_3$ — calcite and aragonite, or $KAlSi_3O_8$ — feldspar, orthoclase). We must be aware, however, that such simple formulae do not always correspond to reality, since minerals include admixtures, as shown in the example of galena.

When the formula of a mineral has been established, it is possible to calculate what percentage of the various constituents should theoretically be present. Every element has a certain atomic weight.* It is necessary to complete the chemical formula with the appropriate numbers and establish the percentage.

Using (Fe_3O_4) as the formula for magnetite, the theoretical percentage composition of individual constituents may be calculated as follows: atomic weight of Fe is 55.85, atomic weight of O is 16. The molecular weight** $Fe_3O_4 = 3 \times 55.85 + 4 \times 16 = 167.55 + 64 = 231.55$. From this molecular weight the content of Fe may be calculated by means of the rule of three. Magnetite $Fe_3O_4 = 100\%$, $Fe = 167.55 = X\%$; $X = \dfrac{167.55 \times 100}{231.55} = 72.4\%$.

Magnetite contains 72.4 % of Fe.

According to their structural crystallographic properties minerals are arranged into systems. According to their chemical compositions, they are arranged into nine classes, as follows:

(1) *Elements* form a comparatively small group of minerals, and in nature are rather rare. From 103 known elements only 22 are found in the native state. Up to now, no light metal has been found in nature in the native state since they easily become oxidized. The number of minerals from the group of elements is naturally larger than 22 as some of them form two minerals, e.g. carbon is diamond and graphite.

(2) *Sulphides and similar compounds;* apart from sulphides, other similar compounds of arsenic, antimony, selenium and tellurium are included since they behave in a similar manner to sulphides.

(3) *Haloids* are compounds of metals with some of the so-called haloid elements, e.g. chlorine, fluorine, bromine, iodine. In comparison with other mineral groups they are comparatively rare in nature. Only fluorite and rock salt are frequent.

(4) *Oxides* are compounds of oxygen and some metallic or non-metallic elements. They are divided into *anhydrous* (quartz, cassiterite) and *hydrated* (opal, goethite, etc.).

(5) *Carbonates* are divided into *anhydrous* (calcite, siderite) and *hydrated* (malachite, azurite, etc.).

---

* The atomic weight of an element is the weight of its atom compared with the weight of an atom of oxygen, taken as 16.

** The molecular weight of a substance is the sum of the atomic weights of the atoms composing a molecule of the substance.

(6) *Sulphates* are also divided into *anhydrous* (anhydrite, barytes) and *hydrated* (gypsum). Closely related to sulphates are *chromates*, *tungstates* (wolframite) and the rare *molybdates*.

(7) *Phosphates* include also the *arsenates* and the rare *vanadates*. Apart from silicates, phosphates are the largest group as regards the number of their kinds.

(8) *Silicates* are the largest group of all minerals, representing approximately 40 % of all mineral sorts. In some silicates the silicon is partly replaced by aluminium. Such silicates are called *aluminosilicates* (e.g. feldspars, kaolinite, etc.).

(9) *Organic compounds* are the following: (a) natural salts of organic acids; (b) natural hydrocarbons, i.e. compounds of carbon and hydrogen in different regular proportions; (c) fossil resins (resins from extinct trees). Also all kinds of coal, crude petroleum, mineral wax and some other substances are sometimes included among minerals.

The number of minerals in individual mineral classes varies considerably. From the total number of 103 chemical elements more than two million compounds of different composition have been prepared in the laboratory. On the other hand, only some 1700 chemical compounds are known as minerals. This is due to the fact that in nature only some chemical compounds are stable, and many rare elements occur only in such small quantities as to be found only as admixtures.

In conclusion to this chapter the main principles of mineralogical terminology should be mentioned. It is quite different from the terminology used in zoology or botany. In contrast to the latter, it is not binominal (i.e. employing two names — those of genus and species), but monominal, using one name for one mineral (usually of international application). The names are most often derived from Greek and are provided with the ending 'ite'. Less frequent are names of Latin origin with endings 'ine' and "ane". Only the most common minerals, familiar from their technical application, are given national names which either become international after some time (passing from one language into all others), e.g. corundum, chalcedony, etc., or differ in different languages. Many minerals' names are derived from proper names, i.e. geographical names, especially the names of various sources of origin (e.g. bauxite after Les Baux near Arles, southern France), or names of famous personalities, especially mineralogists (e.g. goethite).

In the pictorial part of this book, minerals are not arranged according to any system but according to their practical applicability. First of all, mineral ores (ores of iron, manganese, nickel, chromium, tungsten, molybdenum, tin, lead, zinc, copper, silver, mercury, gold, antimony, arsenic and uranium), followed by other applied minerals (raw materials applied in chemical industry, heat resisting raw materials, raw materials of ceramic and glass industries, of building industry, and finally precious stones). At its end the pictorial part contains two interesting groups of rock-forming minerals (amphiboles and augites).

Magnetite, Magnetic Iron Ore

This is the richest iron ore containing 72 % of iron. Magnetite is iron-black, of metallic lustre, and occurs granular or massive. It rarely crystallizes in the form of black octahedra. It is strongly magnetic and most often occurs in skarns, igneous and metamorphic rocks. The most important and richest deposits are considered to be the result of magnetic segregation, as in northern Sweden, Norway, the Urals (USSR) and many other places, such as the USA (Minnesota, Newcomb, New York State, etc.).

In the past, because of its magnetism, it has attracted the attention of many natural scientists and philosophers. It was mentioned by the famous Roman naturalist Pliny the Elder (23—79 A.D.), who described a hill composed of a 'stone' attracting iron, in the vicinity of the river Indus. According to preserved records, however, magnetism was already known by the Chinese as early as the 11th century B.C. It is one of the most valuable ores of iron because of its high iron content, which is also of the best quality. It is easily workable.

Ferro-ferrous oxide — Fe_3O_4, cubic.
Hardness: 5.5—6.5.
Sp. gr.: 4.9—5.2
Streak: black.

1 — magnetite crystal in chloritic schists (Pfitsch, Tirol), 2 — massive magnetite (New Jersey, USA)

Haematite, Iron Glance, Ferric Oxide

It is a very common mineral, causing the reddish colouring of soil, rocks and whole reefs. It contains approximately 70 % metallic iron. In nature it occurs in different forms. Its colour varies, the most frequent shades being dark red to black. The colour of the streak, however, is always blood-red to red-brown, which typifies haematite. Its name stems from the Greek word 'haima', meaning blood.

The most common kinds range from the fine-grained to earthy *red ochre*. The *oolitic* or *fossil iron ore*, formed by sedimentation on floors of former seas, is of great practical importance. Haematite forming globular to reniform aggregates — resembling small skulls — is known as *kidney ore*. *Micaceous haematite* includes the foliaceous and micaceous forms, and in association with quartz composes the rock called *itabirite* (after the occurrence in the Itabira Mountains, Brazil). Beautiful specimens of this come from the island of Elba, where it is found in thick rhombohedral crystals with brightly polished faces, often with a beautiful iridescent tarnish on the surface, which is due to the refraction and dispersal of light by the weathered surface layer of the mineral.

Ferric oxide — Fe_2O_3, hexagonal *Hardness:* 5—6 (with the exception of earthy varieties) *Sp. gr.:* 5.25. *Streak:* red to red-brown.

1 — group of haematite crystals with tarnish colours (Elba), 2 — haematite - Specular Iron (Fichtelgebirge, West Germany) 3 — itabirite (Itabira, Brazil) 4 — haematite crystal

Haematite, Iron Glance, Ferric Oxide

An iridescent variety of haematite is called *specular iron*. It was mentioned as early as Virgil's Aeneid, in which the Roman poet (1st century B.C.) admired its beauty. Heamatite occurs abundantly in many places, the most important deposits being in Sweden, the USA and the USSR.

It is an important iron ore even if the content of iron is lower than in magnetite. It has been known and mined ever since Ancient Times, its compact variety having been used as ornamental stone especially by the Babylonians and Egyptians. The English and Bohemian kidney ores (Horní Blatná near Jáchymov) were used in the past to make stones for sealing-rings and other different intaglios. Red ochre is used as the raw material for making red paint, in polishing and refining gold and silver objects, and in treating certain precious stones.

Ferric oxide — Fe_2O_3, hexagonal. *Hardness:* 5—6 (with the exception of earthy varieties). *Sp. gr.:* 5.25. *Streak:* red to red-brown.

1 — haematite — Kidney Ore (England),
2 — oolitic haematite (Lorraine, France),
3 — massive haematite (Sweden)

1

2

3

Limonite, Brown Haematite, Yellow Iron Ore

It is in fact natural rust containing 48—63 % metallic iron. It is common in occurrence, and is usually the decomposition product of iron minerals whose surfaces it covers with rusty stains. By miners in the past it was truthfully called 'iron hat' (iron cap, ironstone, gossan). Limonite occurs in fibrous, earthy, also compact, reniform or stalactitic forms, often in association with haematite, from which it may be distinguished by the conspicuous yellow-brown colour of its streak. In compact masses its surface sometimes displays an iridescent tarnish.

It occurs extensively in Sweden and Finland, where it is called *bog iron ore* or *lake ore* since it is found on the floors of some lakes; these deposits are caused by the action of bacterial organisms and they are continuously regenerated. On account of its low content of iron it becomes economically valuable only in places of extensive occurrence or where it is associated with other iron ores.

Hydrated iron oxide with alternating content of water — $FeO(OH) + n\,H_2O$, colloidal.
Hardness: variable, most often 4—5.5 (compact 1).
Sp. gr.: 2.7—4.3.
Streak: yellow-brown.

1 — stalactitic limonite (southern Slovakia),
2 — massive limonite — Bog Iron Ore (Finland),
3 — limonite — Bog Iron Ore (Sweden)

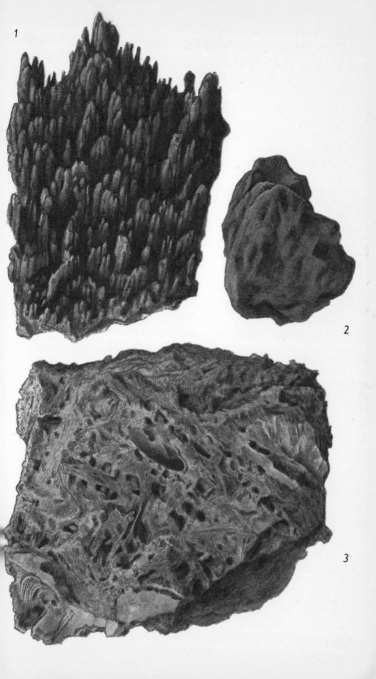

Goethite

It resembles limonite in colour and composition. However, it commonly only forms fine, needle-shaped crystals. Its rusty-brown variety found in Příbram, Bohemia, displays soft, densely growing, needle-shaped crystals of velvety surface. Hence its name 'velvet ore'. Goethite, as well as limonite, may be the result of the weathering of pyrite, or may be formed from hot solutions (hydrothermal origin), or by the dehydration of limonite. The best known deposits — apart from those of Příbram — are found in Siegerland, West Germany, and the Slovak Ore Mountains, where it occurs simultaneously with limonite.

Its economic value is comparatively low; working pays only if it occurs in association with some other iron ores. Because of its beautiful appearance, however, it is much sought after by mineral collectors. Specimens of 'velvet ore' coming from the old abandoned ore mines in Příbram may be found in the larger collections of minerals all over the world.

Hydrated iron oxide — FeO(OH), orthorhombic. *Hardness:* 5—5.5. *Sp. gr.:* 4.3 (decreasing in aggregates to 3.3) *Streak:* brown to yellow-brown.

1 — columnar goethite with quartz crystals (Giant Mountains, Bohemia), 2 — goethite — 'velvet ore' with small crystals of pyrite (Příbram, Bohemia)

1

2

Siderite, Spathic Iron, Chalybite

A minor ore of iron containing 48 % of metallic iron, it is usually brown in colour, massive as well as crystallized. The most perfect rhombohedral crystals, clustered in neat rose-shaped forms and druses, occur most often in veins. Such occurrences, however, are of low economic value since the one contains many impurities. The best example of a siderite deposit is the hill called Erzberg near Eisenerz, Austria, which is entirely composed of siderite. It has been worked there since the 8th century. Originally the hill was composed of limestone, dissolved later by hot solutions rising from the interior of the Earth. In its place siderite was then deposited (hydrothermal metasomatism).

When mixed with clay, sand and organic matter, it is often called *clay ironstone*. It forms dark grey balls and nodules in the coal basins of many countries. They are often nearly half a meter long and very heavy, representing a constant danger to miners (northern England; Halle, German Democratic Republic).

In spite of a rather low content of iron, siderite is an important iron ore. Compared with other iron ores it usually occurs in a relatively pure form which is easy to work.

Ferrous carbonate — $FeCO_3$, hexagonal.
Hardness: 4.
Sp. gr.: 3 (approx).
Streak: colourless.

1 — granular siderite with tarnish colours (Eisenerz, Styria) 2 — group of siderite crystals (Lobenstein, Thuringia), 3 — ironstone balls (northern England)

2

4

Niccolite, Kupfernickel, Copper Nickel, Arsenical Nickel

When Saxon miners of old came across niccolite for the first time while extracting copper ore, they naturally took it for another copper, and tried to obtain copper from it, of course without success. Therefore they called the new ore 'Kupfernickel' — copper nickel (Nickel meaning a rogue in German). Only later was it discovered that the mineral is in fact a compound of arsenic and a new, still unknown element — named nickel. Niccolite contains almost 44 % nickel. It forms fine-grained, yellow-red aggregations of metallic lustre in ore veins and often occurs associated with more valuable cobalt ores. Its surface is often coated with the pale-green *nickel bloom (annabergite)* — hydrous nickel arsenate resulting from the decomposition of nickel minerals. The most important deposits are in La Rioja province, Argentina; in Europe in the Saxon Ore Mountains (Schneeberg), and in Jáchymov and Příbram, Bohemia.

Niccolite is an important nickel ore, applied — like manganese — in ore processing, e.g. in the production of special steels. Since it is very resistant to corrosive agents, it is also used in the manufacture of different equipment, in nickel-coating by galvanism, and in storage batteries. Nickel oxide is added to enamels.

Nickel arsenide - NiAs, hexagonal
Hardness: 5—5.5
Streak: black.

1 — niccolite (Schladming, Styria),
2 — annabergite coatings on niccolite (Jáchymov, Bohemia)

Garnierite (Noumeite); Pimelite

These two closely related minerals are another important source of nickel with their content of 15—33 % of this metal. They most often occur massive or earthy, rarely reniform. Their colour is green, from blue-green through pale apple-green to nearly white. They originated in the course of alteration of olivine rocks due to the removal of the isomorphous admixture of nickel from the olivines, and some of the pyroxenes. Together with other hydrated silicates, they originate and become concentrated in coatings and veinlets in decomposed serpentines. The most important deposits are at Nouméa, New Caledonia (10 % of the world production of nickel ores), and in Cuba; in Europe in Greece, the Polish part of Silesia (Zabkowice Sl.), and the USSR (Khalilovo in the southern Urals and Revda near Sverdlovsk) these minerals occur associated with chromium ores. The small deposits of hydrated nickel silicates composed predominantly of pimelite in serpentine masses in southern Bohemia (Kremže pod Kletí in the area of Český Krumlov) are of a similar type.

Variable hydrate nickel magnesiun silicates, cryptocrystalline
Hardness:
2.5—3.5.
Streak: colourless

1 — garnierite
(Nouméa, New Caledonia),
2 — pimelite
(Nouméa, New Caledonia)

1

2

Chromite, Chromic Iron, Chrome Iron Ore

Chromite is in fact the only chromium ore that contains up to 46 % chromium. It so resembles magnetite as to be almost indistinguishable from it, but the colour of its streak is brown (magnetite has a black streak). It most often occurs as small grains or fine to coarse-grained segregations in olivine rocks, of which it forms the basic component; it also occurs in serpentines, which originated as a result of its decomposition. The chief producers are Rhodesia, the Republic of South Africa, Turkey, the Philippines, and the USSR. It is also found in the USA, Yugoslavia and Albania. It may be of interest that chromium was first discovered in a quite different, comparatively rare mineral called crocoite (crocoisite, crocoise) — lead chromate (PbCrO), forming nice hyacinth-red crystals in deposits of chromite — in Dundas on the island of Tasmania (in Australia) and in Berezovsko in the Urals. Chromium is another element used for the processing of iron. Almost half the world's production is used in the manufacture of special steels of first-rate quality, for use in the manufacture of wagons, bridge frames, heavy arms, and in the aircraft and shipping industries.

Chromic ferrous oxide — $FeCr_2O_4$, cubic.
Hardness: 5.5.
Sp. gr.: 4.5—4.8.
Streak: brown.

Coarse-grained chromite (the Urals, USSR)

Wolframite, Wolfram

Many ores — which are important raw materials at the present time — were formerly considered gangue materials, e.g. sphalerite and nickel ores. Similarly, tungsten ores were formerly considered an unwelcome admixture in melting tin ores. Wolframite is said to have got its name from German miners since in the process of melting, it 'swallowed' (wolf) the tinstone (cassiterite), with which it usually occurs. For a long time the separation of these two minerals was a matter of some difficulty. Also, the element tungsten (wolfram) — discovered in 1781 — was named after wolframite. Black, metallic crystals ranging from columnar to tabular in shape, or perfectly cleavable, scaly and columnar aggregates of wolframite occur most often in tinstone veins, which they often dominate. The most important deposits are in the USA (Boulder Co., Colorado), in Bolivia, in South Korea; in Europe in Portugal, in Saxony and in Bohemia (Cínovec, Horní Slavkov).

Wolfram is predominantly used in the steel industry, in the manufacture of steels for high-speed tools and as filament in electric bulbs, on account of its extremely high melting point — 3390 °C (highest of all metals). Its diameter may be as little as 0.01 millimetre.

Tungstate of iron and manganese — $(Fe,Mn)WO_4$, monoclinic.
Hardness:
4.5—5.5.
Sp. gr.: 7—7.5.
Streak:
from red to brown.

Crystallized wolframite (Cínovec, Bohemia)

Molybdenite

Its name was formed in error from the Greek *molybdaena*, meaning graphite, for which it has often been mistaken. Molybdenite is the most important ore in the production of molybdenum, of which it contains 60 %. It forms perfectly cleavable, metal-grey, hexagonal, flexible scales and laminae which are sometimes very similar to graphite. It differs from graphite in its blue to violet tint and metallic lustre. It occurs mainly in hydrothermally metamorphosed granites and pegmatites, where it is usually associated with tinstone and wolframite. Almost 90 % of world production is supplied from the Climax deposit, Colorado (USA). There are other important deposits in Chile and Quebec, Canada. The richest European deposits are in Yugoslavia. There are also well-known deposits in New South Wales, Australia, and in Malaya, Korea and eastern China.

The principal use of molybdenum is in the manufacture of special steels. Its carbon alloy yields a very hard carbide. It is used in the manufacture of holders of incandescent wolfram filaments in bulbs, also in X-ray lamps and the manufacture of electric furnaces.

Molybdenum sulphide — MoS_2, hexagonal.
Hardness: 1—1.5.
Sp. gr.: 4.7—4.8.
Streak: green-grey.

1 — molybdenite crystal (Renfrew, Canada),
2 — molybdenite scales in quartz (Horní Slavkov, Bohemia),
3 — molybdenite crystal

1
3
2

Cassiterite, Kassiterite, Tinstone, Tin Stone

It is the main and the richest tin ore, containing 78.6 % tin. It forms perfect columnar, black, metallic, tetragonal crystals, often twinned. More often in occurs massive, granular, fibrous or disseminated in rocks (greisen). Associated with wolfram and molybdenum ores, it often forms lodes. It may be also found in alluvial placers. The largest alluvial deposits are in Malaya (Kuala Lumpur), which comprise one third of the world's production, and in Indonesia (Bangka and Belitung islands). The largest primary deposits are in Bolivia and in the Far East in the USSR. In Europe, the chief deposits are in Cornwall, England; Brittany (La Villeder), France; and Bohemia (Ore Mountains).

Tin was one of the first metals used by Man because it can be easily worked and extracted. Man melted it as early as the 6th millenium B.C., before he learned how to melt iron. Tinstone often occurs associated with copper ores; the alloy of these metals is bronze. At the present time tin is used extensively in the manufacture of tin plate (tins, tin foil) for the preservation of food, since it is harmless to human health.

Tin dioxide — SnO_2, tetragonal. *Hardness:* 6—7. *Sp. gr.:* 6.8—7.1. *Streak:* colourless.

1 — twinned cassiterite in quartz (Cínovec, Bohemia), 2 — rounded boulders of cassiterite (Malaya), 3 — twinned cassiterite

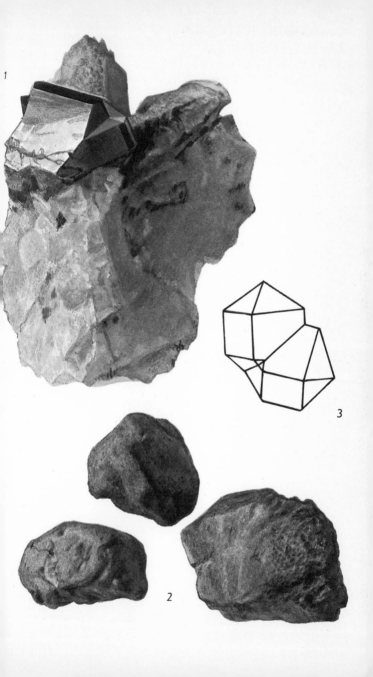

Galena, Lead Glance, Blue Lead

Galena is the chief source of lead, containing 87 % of this metal. It is also one of the most important sources of silver, of which it contains up to 0.5 % (isomorphous admixture). It is often present in veins of different composition, occurring in the form of conspicuous grey-black, metallic granular particles or cubic crystals. It is perfectly cleavable — when struck with a hammer it readily crumbles into small cubes. It is conspicuously heavy. The largest deposits are in the valleys of the Missouri and Mississippi rivers, USA (Joplin Co.), in Europe in Polish Upper Silesia (Bytom); Příbram, Bohemia; Bleiberg, Carinthia; Rodna, Transylvania (Romania), and Broken Hill, Australia.

Galena was first mentioned long ago, in reports by natural scientists in Ancient Times. Lead has been used ever since, because of the ease with which it can be extracted from its compounds. The ancient Romans applied lead in the manufacture of water piping. The most important application, however, after the discovery of print, was as the metal used in type. At present, it is predominantly used for cable-covers and accumulator sheets, to prevent different sorts of irradiation. Lead compounds are employed extensively as pigments, such as those in white and red paints, and in glass- and enamel-making.

Lead sulphide — PbS, cubic.
Hardness: 2.5.
Sp. gr.: 7.3—7.6.
Streak: grey.

1 — group of galena crystals (Joplin, Missouri, USA),
2 — group of galena crystals (Freiberg, Saxony),
3 — cleavage forms of galena,
4 — galena crystal

1

3

4

2

Sphalerite,
Zinc Blende, Blende, Black Jack;
Wurtzite

Both these minerals are of identical chemical composition, differing in their internal structure and in the form of their crystals (polymorphism). They contain 55—67 % zinc. The more common sphalerite got its name from the Greek *sfaleros* (misleading) because of its 'misleading' colour, which varies greatly. It is most commonly brown, black, greenish or reddish, and granular. In fissures it forms perfectly cleavable crystals of a strong metallic lustre. The relatively rare wurtzite is usually brown to yellow-brown, reniform, disk-shaped, stellate. Both minerals occur in ore veins, usually associated with galena (polymetallic ores). Sphalerite is one of the most common minerals. The largest deposits are along the upper course of the Mississippi (USA), in Europe along the frontier between Belgium and Germany. Wurtzite is found in Příbram, Bohemia.

Sphalerite is the chief source of zinc, which is used in large quantities in the manufacture of oxidation-resistant covers and sheets. It is one of the chief sources of gallium, indium and germanium (isomorphous admixtures), which possess excellent properties of semi-conductors. Hence, they are applied in the communications industry in the manufacture of detectors and transistors.

Zinc sulphides — ZnS,
sphalerite cubic;
wurtzite hexagonal.
Hardness:
sphalerite 3—4;
wurtzite 4.
Sp. gr.: sphalerite 3.5—4.2;
wurtzite 3.9—4.1
Streak: sphalerite yellow to brown;
wurtzite brown.

1 — group of dark sphalerite crystals (Derbyshire, England),
2 — group of light sphalerite crystals (Capnic, Romania),
3 — stellate wurtzite (Příbram, Bohemia),
4 — sphalerite crystal

2

4

3

Copper

This is one of the few metals that occur in nature, not only in compounds but also in the native state. Native copper is the richest copper ore, generally containing up to 100 % pure copper. It mostly forms crystalline arborescent forms of a red-brown colour. In common with all minerals containing copper, the surface of native copper displays a bright coloration. It is usually coated with blue azurite and green malachite. In nature it originates by crystallization from hot solutions, or by the decomposition of copper sulphides in the upper part of copper lodes (cementation zone). The name *cuprum* was given to copper by the ancient Romans, after its deposits on the island of Cyprus. The most important locality for the occurrence of native copper is the peninsula of Keewenaw Point on the southern shore of Lake Superior, Michigan; in Europe in Mednorudansk (the Urals) and Vareš (Bosnia).

Copper, as well as tin, was perhaps the first metal used by Man, owing to its extensive occurrence and the ease with which it can be extracted by melting. At present, apart from iron, copper is the most extensively applied metal. It is used above all in the manufacture of alloys (bronze and brass) and in electrotechnology.

Copper — Cu, cubic.
Hardness: 2.5—3.
Sp. gr.: 8.4—8.9.
Streak: red.

Group of copper crystals (Lake Superior, Michigan, U.S.A.)

Chalcopyrite, Copper Pyrites, Yellow Copper Ore

This contains almost 35 % copper. Earlier miners knew it by the iridescent mixture of colours on its surface. These so-called tarnish colours, predominantly violet, blue and reddish, are due to the refraction and breaking up of light rays in the weathered surface layer of the mineral. The colour of a fresh specimen resembles that of pyrite, only it is brighter: brass-yellow to gold-yellow with a slight greenish shade. Chalcopyrite commonly occurs as granular up to massive, vein-filling, rarely in crystals resembling small tetrahedra. It is most often found in ore veins associated with other sulphides, such as galena, sphalerite, pyrrhotite and pyrite. The best known deposits are in the Alps (Mitterberg in Salzkammergut); Mannsfeld, Germany (so-called Kupferschiefer of sedimentary origin); Cornwall, England; Slovak Ore Mountains, Czechoslovakia; French Creek, Pennsylvania (USA); and Arakawa, Japan. Chalcopyrite is the most common copper ore.

Sulphide of copper and iron — $CuFeS_2$, tetragonal.
Hardness: 3—4.
Sp. gr.: 4.1—4.3.
Streak: black-green.

1 — chalcopyrite crystals with quartz (Banská Štiavnica Slovakia),
2 — massive chalcopyrite (Slovak Ore Mountains),
3 — chalcopyrite crystal

3

Malachite,
Green Carbonate of Copper;
Azurite, Chessylite,
Blue Carbonate of Copper

Malachite contains approximately 57%, and azurite 55%, of copper. Both these minerals, which are of similar chemical composition, are common alteration products of copper minerals, causing the typical surface colouring of the latter. Malachite has a conspicuous green colour, different shades of which often follow a concentrically banded arrangement. It usually forms thick reniform encrustations, layers or crystals. The crystals of azurite are of deep azure blue colour; its granular, earthy or pulverulent covers are coloured light blue. The richest deposits of these minerals are in the Urals, in Zaire, in Tsumeb, South-West Africa; Cooper Queen Mine at Bisbee, Arizona (USA); Burra Burra mine near Adelaide, Australia; and in Europe at Chessy near Lyon, France (azurite); Baita and Moldova, Romania; and Siegerland, West Germany. Malachite, especially, is a valuable ore of copper when it occurs in sufficient quantities. Azurite is not so common, and is only rarely extracted. Malachite is often used in jewellery and for ornamental purposes, whereas azurite is seldom applied to this end, being too soft. Ground azurite is used as a blue pigment and in the manufacture of blue vitriol.

Basic carbonates of copper:
malachite — $Cu_2(OH)_2CO_3$;
azurite — $Cu_3(OH)_2(CO_3)_2$
monoclinic.
Hardness: 3.5—4.
Sp. gr.: malachite 3.9—4.03; azurite 3.77—3.89.
Streak: malachite green azurite blue.

1 — crystallized azurite (Chessy, France)
2 — malachite (the Urals, USSR)

Silver

In nature it occurs both as a compound and in the free state. Native silver occurs usually as the alteration product of silver sulphides, e.g. galena and argentite, in the upper part of deposits. It forms black, exceptionally bright silver-white wires. The oldest silver mines in Europe were in Hispania (today Spain), where it was extracted by the Phoenicians, Carthaginians and Moors. Well-known mines were in Freiberg, Saxony, and in Bohemia, where silver has been mined ever since the 7th century. In Bohemia, silver mining was at its prime at the end of the 13th and at the beginning of the 14th century. At that time the mining centre was in Kutná Hora. In the 16th century it was in Jáchymov. Another important locality in Europe is Kongsberg, Norway; outside Europe important silver mines are in Mexico and North America (Lake Superior).

Silver has been used since Ancient Times. In the vaults of the Chaldean kings, dating from as early as the 4th millenium B.C., archaeologists found jewels and objects made of silver. Native silver is too soft to be of practical use in the pure state. Therefore it is used in alloys with other argentiferous ores, in coinage and jewellery; and as silver salts in medicine and photography.

Silver — Ag, cubic.
Hardness: 2.5.
Sp. gr.: 10—12 native, 10.5 pure.
Streak: white.

Silver wires
(Kongsberg, Norway)

Argentite, Silver Glance

This contains more than 87 % silver. It occurs as massive filling of veins or as arborescent pseudomorphs after native silver. Less often it crystallizes, with imperfect, rounded crystals of irregular growth. Fresh surfaces are light grey, of high metallic lustre resembling galena. On exposure it soon becomes dull and black and it may be scratched and cut by a knife. It occurs commonly in veins associated with other silver ores, especially galena. The most important localities are Freiberg and Schneeberg, Saxony; Jáchymov and Banská Štiavnica, Czechoslovakia; and the famous Comstock Lode in Nevada (USA).

Apart from native silver, it is the richest silver ore, mentioned in much detail already by one of the founders of mineralogy, Georgius Agricola, Georg Bauer by his own name (1494—1555), who lived in Saxony and Joachimsthal (now Jáchymov), Bohemia.

Silver sulphide — Ag_2S, orthorhombic-pseudocubic.
Hardness: 2.
Sp. gr.: 7.3.
Streak: grey to black.

Group of argenti crystals (Banská Štiavnica, Slovakia)

Cinnabar, Natural Vermilion

This contains 86 % mercury and occurs as compact, massive or earthy aggregates and crusts of a cochineal red colour. Crystals are rare. It is found in veins, especially in areas of surface weathering in the so-called iron hats (gossans, iron caps). The ancient Romans worked cinnabar as early as the 7th century B.C. in southern Spain, especially in the vicinity of Almaden, which is still today one of the richest deposits of mercury. Other important localities are Monte Amiata in Tuscany, Italy, and Idria, Slovenia and Avala near Belgrade, Yugoslavia. Outside Europe, the most important deposits are in the USA, in Texas, Oregon, and California; in Mexico; Hunan and Szechwan provinces in China; in the Caucasus (USSR); and Khaydarken and Chauvay, Central Asia.

Cinnabar is the chief source of metallic mercury, which was originally applied in the recovery of silver and gold from ores, and in the manufacture of mirrors. It is interesting that the consumption of mercury has not increased in the last decades, as it has successfully been replaced by cheaper materials. It is used especially in medicine, in the manufacture of ammunitions and in the purification of valuable metals. Mercury is also used in the manufacture of paints.

Mercuric sulphide
— HgS,
hexagonal.
Hardness: 2.5.
Sp. gr.: 8.1.
Streak: scarlet.

1 — cinnabar crystals (Avala near Belgrade, Yugoslavia),
2 — massive cinnabar (Idria, Slovenia),
3 — cinnabar crystal

1

Gold

Gold is not affected by oxidation, nor does it alter under normal conditions, so that in nature it occurs almost always native, and only exceptionally in combination with other substances. It is usually found disseminated in quartz veins. As the veins become reduced by the action of atmospheric agencies and erosion to sand and gravel, the gold, on account of its high specific gravity, becomes concentrated in secondary deposits of placer or alluvial type. It is generally so finely dispersed as to be almost invisible to the naked eye. Occasionally, it forms larger, yellow, soft scales, grains, wires and nuggets. Up to 50 % of the world's production of gold had been supplied until recently from the old alluvial deposits at Witwatersrand and Odendaalsrus in the Transvaal, South Africa. Similar deposits, but smaller and of less importance, are in the USA, where it occurs in quartz veins in plutonic granites (e.g. the vein system, Mother Lode, California). Similar deposits are also found in the USSR and in France.

It is virtually impossible to think of a metal which has played a more important part in the history of mankind than gold. Since Ancient Times gold has been the symbol of power and riches. It is a valuable metal, used especially in jewellery, medicine, coinage and glass-making, in technology (for contacts and fountain pens), and in dentistry in alloys with other metals.

Gold — Au, cubic.
Hardness: 2.5—3.
Sp. gr.: 19.28 (pure).
Streak: yellow.

Gold sheets between crystals of quartz (Rosia Montana, Romania)

Stibnite, Antimonite, Antimony Glance, Grey Antimony

This contains up to 71 % of antimony and occurs in ore veins associated with other ores, or in isolated, sometimes auriferous, veins. It forms elongated crystals developing into capillary forms. Sometimes compact. Its coloration is lead-grey with a slight bluish tint. The beautiful crystals of stibnite found in the island of Shikoku, Japan, are famous and have been used for ornamental purposes since earliest times. The richest deposits in Europe are in Czechoslovakia (Bohutín near Příbram, Milešov, the Low Carpathians), Italy (Sardinia and Tuscany), and Yugoslavia (near Fojnica, Krupanj and Kostajnik). The most important deposits are in the Republic of South Africa, in Bolivia (Chayanta, Palca and Tupiza regions), in China (Hunan and Yunnan provinces) and Mexico.

The Roman naturalist Pliny described it as a cosmetic preparation much in use in his time. Ground silver-grey antimonite was applied for darkening the eyelids. At present it is the chief source of antimony, three quarters of which are used in the production of different alloys of economic importance. The most important is the metal type applied in letter-press printing (antimony with lead).

Antimony trisulphide — Sb_2S_3, orthorhombic. *Hardness:* 2—2.5. *Sp. gr.:* 4.63. *Streak:* grey.

1 — cluster of stibnite crystals (Shikoku, Japan),
2 — columnar stibnite (Bohutín, Bohemia)

1

2

Arsenopyrite, Mispickel, Arsenical Pyrites

This is a typical lode-ore containing 46 % arsenic, occurring both in association with other sulphur ores (polymetallic veins) and in the pure form. It resembles pyrite because of its yellowish tarnish on exposure. The fresh-fracture surface is of tin-white colour, inclined to steel-grey. When struck, it emits a garlic odour typical of arsenic. It occurs massive, granular, columnar or in rodlike aggregates. The largest deposits are in the Aude Department, France, and in Freiberg, Saxony. In Mexico, the USA and Australia it is extracted from polymetallic veins. Smaller deposits occur in serpentines near Złoty Stok, Silesia (Poland), and in Czechoslovakia (polymetallic veins in the Giant Mountains). Although arsenopyrite contains a comparatively high percentage of arsenic, 90 % of this element is at present recovered as a by-product in the treatment of other ores which contain arsenic only as admixture.

The element arsenic as well as its compounds were known as poisons from earliest times. Today it is extensively used, especially in the production of medicines, dyes and insecticides (in agriculture).

Iron sulpharsenide — FeAsS, orthorhombic.
Hardness: 5.5.
Sp. gr.: 6.07—6.15.
Streak: black.

Crystallized arsenopyrite (Freiberg, Saxony)

Uraninite, Pitchblende

In 1896, the French physicist Henri Becquerel discovered that uraninite emitted special radiation which he called radioactivity. Shortly afterwards, Maria Curie-Sklodowska discovered in uraninite the elements radium and polonium, which give more irradiation than uranium. Uraninite contains approximately 80—90 % uranium, and radium and polonium in trace amounts. It was called pitchblende because of its close resemblance to pitch. It generally occurs as compact vein filling, less often in reniform aggregates. If placed with a cut face on a photographic plate, the latter becomes irradiated. Uraninite occurs in ore veins, pegmatites, granites and coals. For their experiments, both Becquerel and Maria Curie-Sklodowska used specimens of pitchblende from Jáchymov, Bohemia. The largest deposits of uraninite are in Canada, north of Lake Huron, and in Zaire (Katanga, Chingolobwe).

Uraninite, as the most important uranium ore and chief source of other radioactive elements, has become the subject of world-wide interest in the age of harnessing atomic energy both for peaceful and military purposes. Thus, it has become the most important strategical raw material. Radium — used especially in medicine — occurs in uraninite in very small quantities, e.g. from a wagon of uraninite ore less than 1 gr of radium can be recovered.

Oxide of uranium with admixture of thorium — (U,Th)O$_2$, cubic. *Hardness:* 5—6. *Sp. gr.:* 8—10. *Streak:* red-brown

1 — imperfect crystal of uraninite (Chingolobwe, Zaire), 2 — reniform uraninite (Jáchymov, Bohemia)

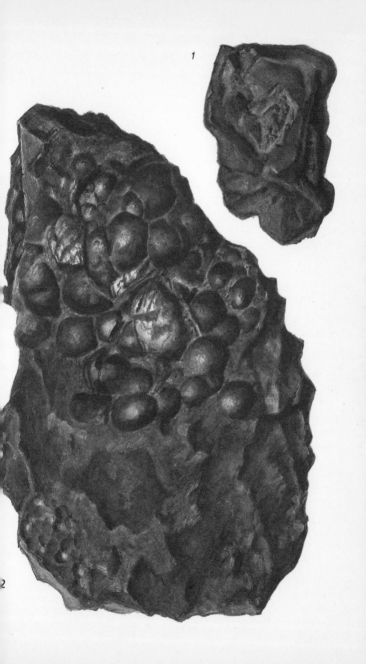

Bauxite

The content of aluminium in bauxites varies from 25 to 30 %. Bauxite occurs commonly in earthy, granular or pisolitic masses of dirty white-grey, yellowish, brown to dark red-brown colour. It is a decomposition product of feldspathic rocks, or the result of tropical weathering. It is named after its original source: Les Baux, near Arles, southern France. Large deposits are also found in Yugoslavia and Hungary, the most important deposits occurring in Jamaica, the Netherlands and Guyana.

Bauxite is the chief industrial source of aluminium, which is rightly called the metal of the 20th century. Aluminium is one of the latest metals which have been successfully produced artificially. Small amounts of aluminium were produced 45 years ago but at that time no one thought it could have any social significance because its extraction was too difficult. Much later, however, it was successfully produced from some of its compounds by electrolysis. Since then the world's production of aluminium has continuously been rising. At present it is mainly used in the aircraft, automobile and shipping industries, and also in fuel elements for atom reactors.

Sedimentary rock whose main components are hydrated aluminium oxides.
Streak: white.

Bauxite
(Les Baux, France)

Sulphur

There are many minerals containing sulphur in nature. It occurs both in the native state and in combination. It was one of the first elements recognized by ancient Man and because it was found in the craters of extinct volcanoes and could be burnt, supernatural powers were ascribed to it. Apart from deposits of volcanic origin, however, sulphur is also formed by the action of hot springs (by which it is deposited with tufa), and by the action of organisms. Almost everywhere it originates by sublimation in fires of coal heaps. It forms light yellow or honey, yellow-brown crystals, compact masses and encrustations. In ancient times the Romans knew of sulphur from the southern Sicilian deposits, which have been yielding considerable quantities up to the present day. Important deposits exist in the vicinity of Tarnobrzeg, Poland. The world deposits amount to $\frac{1}{4}$ billion tons, the most important producer being the USA (75 % — Louisiana and Texas with the Boling deposit). Deposits of free sulphur are especially important since sulphur is recovered by mere purification from the associated gangue by melting in ovens.

Sulphur has become indispensable for the chemical industry and for a number of other branches of industry it is most important. There are only a few manufacturing processes which can do without it.

Sulphur — S, orthorhombic. *Hardness:* 2. *Sp. gr.:* 2. *Streak:* colourless to yellow.

Crystals of sulphur (southern Sicily)

Pyrite, Iron Pyrites;
Marcasite, White Iron Pyrites

Both are of the same chemical composition, differing, however, in internal structure and shape of their crystals. They contain approximately 54 % of sulphur and 46 % of iron. Pyrite forms golden-yellow grains and crystals. Typical forms are striated cubes, octahedra and pyritohedra. Crystals of marcasite are commonly small, columnar, often twinned, and are of a bronze-yellow colour, paler than pyrite. Both occur associated in coals. The largest deposits occur in unaltered, as well as metamorphic, sedimentary rocks as a result of decomposition of vegetable and animal remains. The most perfect crystals are found in ore veins. The chief producers are Spain (Rio Tinto) and Japan; the best known crystals come from Elba. Marcasite is not as abundant as pyrite. It occurs in the chalk marl at Folkestone, southern England, and in Freiberg, Saxony.

Pyrite is an important raw material of economic value. It is especially applied in the manufacture of sulphuric acid, sulphates, sulphur, paints and polishes. It is often treated by roasting. The refuse may be used in smelting processes or as iron ores of inferior quality. Some pyrites contain gold and copper.

Iron bisulphide - FeS_2, pyrite cubic; marcasite orthorhombic. *Hardness:* pyrite 6—6.5; marcasite 5—6. *Sp. gr.:* pyrite 4.9—5.2; marcasite 4.6—4.9. *Streak:* pyrite brown-black; marcasite green-grey.

1 — crystal of pyrite (Elba, Italy), 2 — crystallized marcasite (Most, Bohemia 3 — crystal of pyrite — deltoid dodecahedron, 4 — crystal of pyrite — striate cube

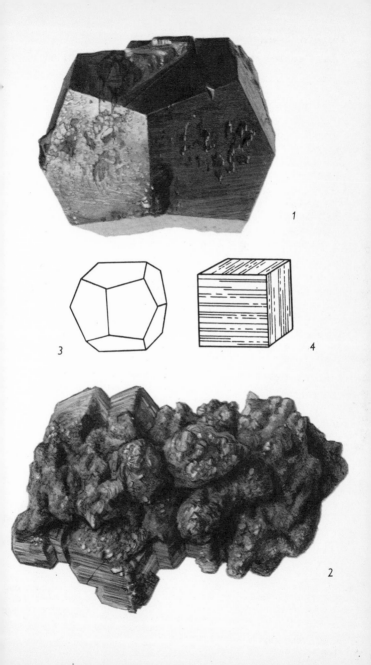

1
3
4
2

Barytes, Barite, Heavy Spar

Barytes was familiar even to the first miners since it is a common and widely distributed mineral, occurring in cavities of ore veins associated with sulphide ores; especially lead, zinc, and silver. It was formed by deposition from hot solutions rising from the interior of the Earth at higher temperatures. It sometimes occurs in isolated veins or in sedimentary deposits. Distinctive is its specific gravity (hence its name Schwerspat in German, and heavy spar in English). It often forms beautiful crystals, comparatively large and thick orthorhombic plates and columns. Compact barytes forms different aggregates. It is usually of a light colour. The chief world producers are the USA, Germany (the Meggen deposit), and England.

It had long been considered a worthless gangue material, although in the Middle Ages it aroused much excitement among alchemists. In Bologna, the Italian alchemist Vincenzo Cascatiolo, heating barytes in the course of his experiments, discovered that when heated it phosphoresces in darkness. Thus, its luminescence was discovered. The chemical industry uses barytes primarily in the manufacture of paints, to give weight to paper, etc. In the ceramic industry it is used in the manufacture of glazes and enamels.

Barium sulphate — $BaSO_4$, orthorhombic. *Hardness:* 2.5—3.5. *Sp. gr.:* 4.3—4.6. *Streak:* colourless to white.

Crystals of baryte (Cumberland, England)

Rock Salt, Common Salt, Halite

Salt is present in the waters of the oceans, some 20 million km³ of salt being dissolved in the sea water. There exist, however, extensive inland deposits formed at various geological times by evaporation of salt water in enclosed, or partly enclosed, bodies of sea water in hot and dry climates. Rock salt forms compact to granular masses and cubic crystals in fissures, which are mostly colourless but may sometimes be slightly coloured when impure, especially of bluish tint. Salt has an excellent cubic cleavage; when struck by a hammer it crumbles into many tiny cubes. The chief producers are the USA, Great Britain and the Federal Republic of Germany. The best known deposits, however, are in Austria in the Salzkammergut, and in Poland in the vicinity of Wieliczka and Bochnia.

Salt is indispensable for human life. The average annual consumption per head is 7 kg. Therefore, early in history it was of a considerable commercial importance. Huge quantities are used in the chemical industry, especially in the manufacture of sodium compounds, chlorine, and chlorine compounds. The large pure crystals are used in optical apparatus, and in working with infra-red rays.

Sodium chloride — NaCl, cubic.
Hardness: 2—2.5.
Sp. gr.: 2.1—2.3.
Streak: colourless to white.

1 — crystals of rock salt (Wieliczka, Poland),
2 — fibrous rock salt (Salzkammergut, Austria),
3 — crystal of rock salt (Wieliczka, Poland)

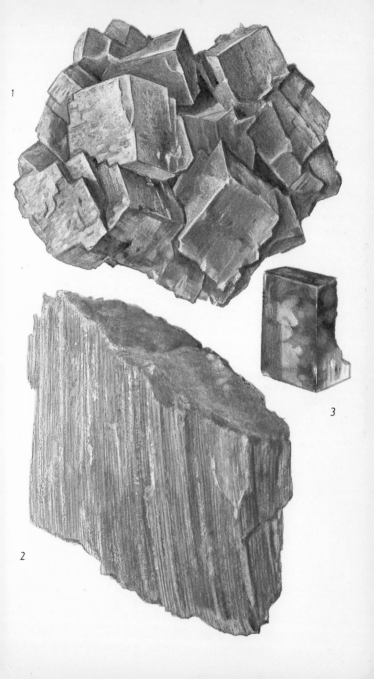

1

2

3

Apatite: Fluor-apatite, Chlor-apatite, Hydroxyl-apatite

The name apatite is derived from the Greek *apatao*, 'I am misleading'. The mineral was given its name by past mineralogists on account of its misleading resemblance to a number of other minerals. Indeed, its crystals occur in many different shapes, from thick plates and columns to long drawn-out needles. It also occurs massive, granular and compact. Its colour varies greatly, from translucent to white, green, and blues of different shades, to violet and brown. It is present in almost all rocks, especially pegmatites and tinstone deposits but only in small quantities. It often occurs in bedded sedimentary deposits. Such deposits are called *phosphate rocks*. The world's largest deposits are in the USA (Florida and Tennessee — yielding almost one half of the world's production of phosphates), North Africa (Tunisia, Morocco, Algeria and Egypt), the USSR (Kazakhstan, Ukraine and the Kola peninsula) and France. The best known apatite crystals come from St. Gotthard (Switzerland) and Quebec (Canada). Phosphates are used in the manufacture of fertilizers.

Calcium fluorphosphate (fluor-apatite); calcium chlorphosphate (chlor-apatite); calcium hydroxyl-phosphate (hydroxyl-apatite) — $Ca_5(F,Cl,OH)(PO_4)_3$, hexagonal.
Hardness: 5.
Sp.gr.: 3.17—3.23.
Streak: white to grey.

1 — crystals of apatite (Horní Slavkov, Bohemia),
2 — phosphorite (Podolia, Ukraine)

1

2

Fluorite, Fluorspar

This is the most common natural compound, and the raw material of the gas fluorine which was named after it. It occurs in pegmatites and similar rocks, and in hydrothermal veins. It is one of the few minerals that occurs comparatively often in crystalline forms and in striking colours. Its cubic crystals are usually dark violet, yellow or green, more rarely red, pink or colourless. It is no wonder that miners of earlier times, in the Ore Mountains, Bohemia, called it 'Ore Bloom'. The largest deposits occur in the USA; in Europe in Saxony (Freiberg), in England (Weardale near Durham and Derbyshire), in France and in the USSR (also in Transbaykalia).

It is an important raw material used as a flux in steel-making. Its colourless variety is used for the construction of lenses transmitting ultraviolet rays. Fluorine and some of its compounds are poisonous and therefore behave as disinfectants in small doses. Other fluorine compounds help to lower the temperature to as low as $- 100\ ^{\circ}C$. On account of this property they are used in can-making and also in the manufacture of plastics. Fluorine is used in separating uranium and its isotopes, and together with other light elements, as a fuel in cosmic rockets.

Calcium fluoride — CaF_2, cubic.
Hardness: 4.
Sp. gr.: 3—3.25.
Streak: white.

1 — fluorite (Annaberg, Saxony),
2 — fluorite (Derbyshire, England),
3 — penetration twins of two fluorite cubes

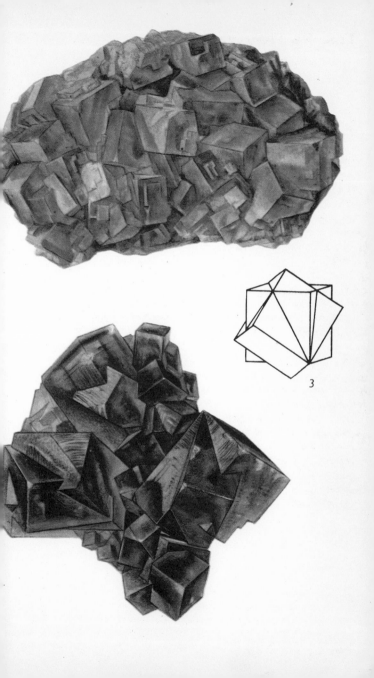

3

Graphite, Black Lead

This is one of the varieties of carbon, another one being diamond. The enormous differences in the properties of these two minerals are due to a different atomic arrangement. In graphite the atoms are arranged in layers which are further apart than in diamond. This also accounts for the differences in the physical properties, e.g. diamond is a hard mineral, whereas graphite is so soft that it smears fingers when touched. Diamond is a rare mineral, graphite is quite common. In nature it is usually found in scales, laminae up to massive bodies, black to black-grey in colour. It originates by the decomposition of carbon dioxide, as a result of volcanic activity or by the metamorphism of carbonaceous material of sedimentary origin. It occurs in metamorphosed rocks, less often in magmatic rocks. Important deposits of graphite are in the Bohemian Forest (Czechoslovakia), in Bavaria and Ceylon.

Graphite was formerly used especially for the manufacture of pencils. At the present time it is chiefly used for crucibles, in the electronic industry, in the manufacture of lubricants, as well as in nuclear technology, especially in reactors, since it effectively slows down neutrons, absorbing them only slightly.

Carbon — C, hexagonal.
Hardness: 1.
Sp. gr.: 2.2.
Streak: black.

Graphite
(Ceylon)

Serpentine

This obtained its name from the Latin word *serpens* (snake) since it was believed to be an effective remedy against snake-bite. Mostly it occurs in the form of *chrysotile*, from the Greek words *chrysos* (golden) and *tilos* (fibre), or *serpentine asbestos*, i.e. fine-fibrous serpentine occurring as small veins in *serpentine rock — serpentinite*. Serpentinite is composed of microscopic fibres of chrysotile and of mineral remains of volcanic rocks by whose decomposition in water it originated. It is translucent to opaque, from yellow almost to black, and has differing shades of green. Asbestos was known in antiquity when it was worked in Cyprus. The chief European deposit is in Balangero near Turin; the richest deposits are in Canada. In central Europe an important deposit of serpentinite is in Zöblitz, Saxony.

Chrysotile, on account of its fine-fibrous structure, provides the most important part of the asbestos of commerce, i.e. the fine-fibrous varieties of some silicates are applied in the production of fire- and heat-resistant cloths, etc. Until the 19th century, objects made of asbestos were considered more as curiosities than as items of practical use. Its practical application was fully recognized only recently, and it is now used for the manufacture of fire-proof fabrics, roofing tiles, boards for insulating purposes in electrotechnology, and different fire-proof paints.

Hydrous magnesium silicate — $Mg_3(OH)_4Si_2O_5$, monoclinic, pseudorhombic. *Hardness:* 3—4. *Sp. gr.:* 2.5—2.7. *Streak:* white.

1 — massive serpentine (Saxony), 2 — chrysotile (Canada)

1

2

Micas:
muscovite, biotite, fuchsite, zinnwaldite, lepidolite, phlogopite

They represent a group of minerals distinguished by a perfect basal cleavage. According to their colour they are divided into *light* and *dark micas*. The most common light mica is *muscovite* (Potash Mica, Common Mica, Muscovy Glass), occurring in the form of large transparent plates, in fine disseminated scales or massive (sericite). Crystals are scarce. The fine-scaly variety is usually white, pearly; the massive variety is greenish, while the lamellar variety is colourless. The finely scaled emerald-green muscovite with the admixture of chromium is called *fuchsite*. The light mica containing lithium and iron is called *zinnwaldite*. The latter has obtained its name from its source in Zinnwald (now Cínovec), on the frontier between Bohemia and Saxony. It is usually lamellar to tabular, of dark silver-grey colour with a bronze or green tinge. *Lepidolite* (Lithia Mica) contains lithium, too. The first known deposit of lepidolite is at Rožná, Moravia. On the verge between light and dark micas is the light red-brown *phlogopite*. This contains fluorine. Of the dark micas — whose colouring is caused by the presence of iron — the most common is *biotite*. It forms generally black, brown and dark green imperfect crystals. Micas are very common. They occur as the main constituent of many rocks, such as pegmatites (muscovite). Consequently, they occur frequently as-

Silicates of aluminium, potassium and hydrogen, with fluorine and some other elements:
muscovite — $KAl_2(OH,F)_2 AlSi_3O_{10}$;
zinnwaldite — $KLiFeAl(F,OH) AlSi_3O_{10}$;
lepidolite — very complex silicate with Li;
biotite — $K(Mg,Fe,Mn)_3 (OH,F)_2AlSi_3O_1$
phlogopite — $KMg_3(F,OH)_2 AlSi_3O_{10}$.

Plate of muscov
(India)

Micas

sociated with feldspars. Dark micas are less resistant, and — in contrast to muscovite — they decompose easily. Mostly they change into greenish chlorides (silicates similar to micas), or into muscovite, due to the loss of iron. Sometimes gold-brown, metallic scales are formed in this way — called *rubelane*. The largest sheets of light mica are found in the Urals, from which they have been imported to western Europe since long ago. On account of this they were called 'Muscovy Glass', which gave rise to the name of muscovite. The current world production of micas is about 50,000 tons per year, the chief producers being the USA (about 50 %) and India (23 %). The North American deposits of muscovite are predominantly in North Carolina. The deposits in St. Gotthard (Switzerland) are particularly mineralogically interesting. Crystallized biotite is found on Vesuvius (Italy) and in Templeton (Canada).

The perfect cleavage and transparency of micas were already made use of as much as 300 years ago. Mica sheets were formerly used as window panes. Today, light micas are of great importance on account of their heat-resistance and because they do not decompose in acids. They are chiefly used as insulating materials in electrotechnology and optics. Because of their low resistance, dark micas are of little use in technology. Ground biotite is sometimes used as a special paint.

Monoclinic-pseudohexagonal
Hardness: 2—3 (lower with light micas, higher with dark varieties).
Sp. gr.: 2.7—3.1.
Streak: white to grey.

1 — imperfect crystal of biotite (Templeton, Canada),
2 — crystals of zinnwaldite (Cínovec, Bohemia)

1

2

Feldspars:

orthoclase, adularia, sanidine,
microcline, amazonstone, albite,
oligoclase, labradorite, anorthite

In the course of their decomposition many minerals provide substances which may become essential constituents in the nutrition of plants. They consist predominantly of silicates which are present in the majority of rocks. The most important among them are feldspars. They got their name from the word 'field' since they enrich arable soil. The most common and most important is *orthoclase* — potash feldspar. Also the soda-lime-feldspars, i.e. the *plagioclases*, are of importance. They form a continuous isomorphous series with soda-feldspar, *albite*, and lime-feldspar, *anorthite*, as end-members. From intermediate members the best known are *oligoclase* (with prevailing soda component) and *labradorite* (with prevailing lime component). Less frequent is *microcline*, another kind of potash feldspar. It crystallizes in quite a different way from orthoclase. Feldspars have a perfect cleavage. They occur compact, granular or lamellar, or form short-tabular or columnar crystals. They show simple twinning on the Carlsbad, Manebach or Baveno laws (named after their sources). The orthoclase variety, forming transparent glassy tabular crystals is called *sanidine*. Potash feldspars are usually white, yellowish or pinkish. The purest variety of orthoclase is called *adularia* (after Mount Adula near St. Gotthard, Switzerland). It is transparent, often with fine tarnish colours. Plagioclases are usually darker, sometimes exhibiting brilliant tarnish colours. The bright-green variety is called *amazonstone*.

Potassium-aluminium or soda-lime silicates: orthoclase — $KAlSi_3O_8$; microcline — $(K,Na)AlSi_3O_8$; albite — $NaAlSi_3O_8$; oligoclase — $(Na,Ca)AlSi_3O_8$; labradorite — $(Ca,Na)AlSi_3O_8$; anorthite — $CaAlSi_3O_8$.

1 — crystal of orthoclase — Carlsbad Twin (vicinity of Karlovy Vary, Bohemia), 2 — crystal of amazonstone (Pikes Peak, Colorado, USA), 3 — Carlsbad Twin

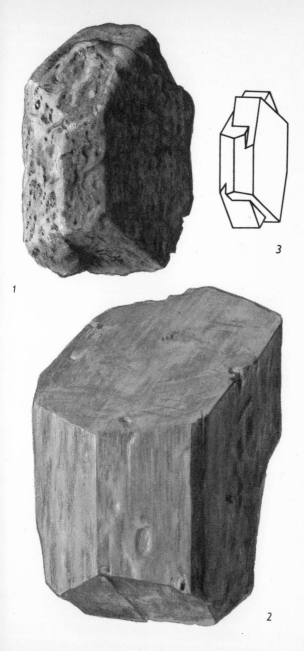

1
2
3

Feldspars

Feldspars are the most important group of rock-forming minerals. They are the dominant components of most igneous rocks, such as granites and pegmatites (especially potash feldspars). Plagioclases occur especially in more basic igneous rocks, such as diorites, melaphyres, porphyrites, basalts and tephrites. Feldspars are also an essential part of nearly all the common kinds, e.g. schists and sedimentary rocks (arkoses). The chief producer of feldspars is the USA (over $\frac{1}{4}$ million tons of feldspathic raw material per year). Deposits occur especially along the coast of the Atlantic Ocean (South and North Carolina, Maine) and in South Dakota. Important deposits are also in Canada, Sweden, Norway, France, the USSR, and elsewhere. Well-known deposits of orthoclase are in Karlovy Vary and Dolní Bory (Bohemia), and Baveno (Italy).

Feldspars, especially orthoclase, are of economic importance. They are used in glass and ceramic industries in the manufacture of porcelain, porcelain-glass and enamels. Some kinds of feldspars are cut for ornamental purposes (especially adularia, amazonstone, and labradorite).

Orthoclase monoclinic; microcline and plagioclase triclinic.
Hardness: orthoclase and microcline 6; plagioclases 6—6.5.
Sp. gr.: orthoclase 2.54—2.56; microcline 2.57; plagioclase 2.62—2.76.
Streak: white to grey.

1 — crystals of adularia (St. Gotthard, Switzerland), 2 — labradorite (Labrador, Canada)

1

2

Kaolin, China Clay

This is a white, soft clayey material whose main constituent is *kaolinite*. Kaolinite is usually lamellar to earthy, with a greasy feel. Apart from kaolin it contains some other aluminium silicates, such as *nacrite* (from the Persian word *nacro*, meaning complexion) and *dickite*. Kaolin often contains admixtures of other minerals. It has resulted from the alteration of aluminium silicates, especially feldspars, in granites and arkoses (fine-grained sediments composed of quartz and feldspar). They occur in large quantities in the Kiangsi province in China. The most important European deposits are in Bohemia (in the vicinity of Karlovy Vary, and in the area of Plzeň near Horní Bříza), in England and the German Democratic Republic.

The name 'kaolin' is of Chinese origin. The Chinese worked it as early as in the 6th century A.D., i.e. ground it, mixed it with water, cleaned by washing, dried the fine deposit, and pressed it into porcelain. They had long kept this manufacturing process a secret. At present only kaolins of the best quality are applied in the manufacture of porcelain. Poorer quality kaolins are used as fillers in paper and paint manufacture, in the manufacture of crockery, and in the chemical industry in the manufacture of ultramarine.

Sedimentary rock composed mainly of hydrated aluminium silicates.
Streak: white to grey.

Kaolin (vicinity of Karlovy Vary, Bohemia)

Calcite, Calc Spar, Carbonate of Lime

Apart from quartz, it is the most common mineral. It occurs in many different forms, from crystals (bounded by perfect faces) to minute microscopic grains building extensive beds of different geological ages. It crystallizes in rhombohedra and scalenohedra of great variety and complexity of form, often arranged in druses. In nature it occurs white, sometimes with different tints due to admixtures. Pure calcite is always colourless. A small rhombohedron of pure calcite is perfectly transparent and gives a high double refraction when one looks directly through it, which is an interesting optical phenomenon. Its discovery was of much help in optical research.

Besides the crystalline forms, calcite occurs in the form of massive, fine- to coarse-grained or compact rocks, sometimes forming extensive beds and long mountain ranges. These rocks have various names. *Travertines,* for instance, are cellular deposits of calcium carbonate derived from waters charged with calcareous matter in solution (they often contain remains of water plants, especially algae). *Marls* and *arenaceous marls* are compact limestones containing dispersed particles of clay and sand. The most important are *marbles* (crystalline limestones), which are often beautifully coloured. These differing varieties of calcite originated under extremely different conditions. Crystallized calcite originates in veins, in caverns of limestone rocks and cavities in various igneous rocks. In the latter it is usually formed by deposition from hot solutions. It may be found almost everywhere.

Calcium carbonate — $CaCO_3$, hexagonal. *Hardness:* 3. *Sp. gr.:* 2.6—2.8. *Streak:* colourless.

1 — travertine (southern Slovakia), 2 — marble (Carrara, Italy)

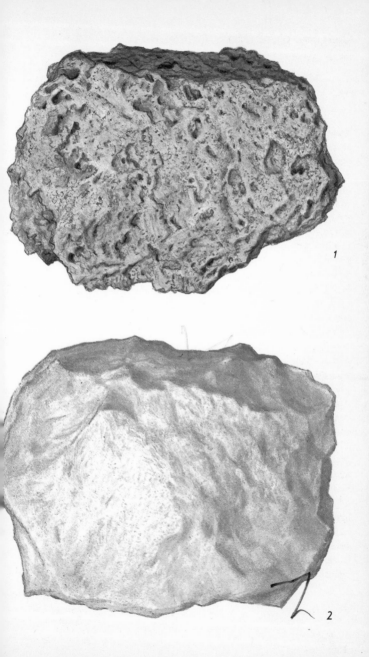

1

2

Calcite, Calc Spar, Carbonate of Lime

Most limestone rocks have been deposited on the sea bottom. Rivers carry large quantities of small particles of calcium carbonate from weathered rocks into seas and oceans. Some animals use it to build their skeletons and shells, the remains of which have accumulated in the vast ocean beds.

The most important deposits of crystallized calcite are at Helgustadire near Eskifjord, Iceland (birefringent calcite), and at Joplin Co., USA. A special variety — much sought after by collectors — is the so-called *nail-head spar*, a characteristic mineral of Příbram, Bohemia. It is formed of parallel twins of flat rhombohedra arranged so that the upper crystals resemble nail heads. Famous deposits of marble are in Greece.

Calcium carbonate — $CaCO_3$, hexagonal. *Hardness:* 3. *Sp. gr.:* 2.6—2.8. *Streak:* colourless.

1 — birefringent limestone (Eskifjord, Iceland), 2 — rhombohedral crystals of calcite (Příbram Bohemia)

1

2

Calcite, Calc Spar, Carbonate of Lime

Marbles from Paros and Pentelikon were used by the ancient Greeks as material for sculpture. Famous marbles also come from Italy, where they were quarried on the slopes of the Apuanian Alps, especially at Carrara as early as in the 13th century. We can still admire the beautiful marble sculptures dating from ancient Greece and Rome. Later, in the Middle Ages, marble was used especially for decorating churches. Since that time it has remained one of the most popular ornamental stones used for sculpture and wall facing. At present, all varieties of calcite have become important raw materials, applied especially in the building industry (quicklime, mortar, cement and concrete). Carbonate of lime finds many different uses according to its purity and character — the varieties containing some clayey matter are burnt for cement. In the chemical industry it is used especially in the manufacture of soda, carbon dioxide, calcium carbides, cellulose, bleaching powder, plaster and fertilizers. Calcium carbonate is also used as a flux in ore smelting and in glass industry. Clear transparent forms of calcite are used in the construction of optical apparatus.

Calcium carbonate — $CaCO_3$, hexagonal. *Hardness:* 3. *Sp. gr.:* 2.6—2.8. *Streak:* colourless.

1 — scalenohedral crystal of calcite (Joplin, USA),
2 — crystals of calcite (Cumberland, England),
3 — scalenohedral crystal of calcite

3

Aragonite

The Swedish natural scientist K. v. Linné discovered that aragonite had the same chemical composition as calcite. Later it was proved, however, that many of its physical properties are different from those of calcite. It crystallizes in the orthorhombic system, and though the forms may occasionally look like those of calcite, they are easily distinguished because they do not show its cleavage. Columnar crystals have often grown into six-sided prisms. Crystals are usually slender needles up to long columns, often arranged in druses. It may be colourless or snow-white, light grey, yellow or pinkish. Apart from crystalline forms, aragonite also occurs compact or granular, e.g. the finely foliated *sprudelstein* and the more valuable variety, *peastone*. Both these varieties are formed by deposition from hot springs, e.g. in Karlovy Vary (formerly Carlsbad), Bohemia. Aragonite, like calcite, originates under variable conditions. Calcite crystallizes from comparatively cool solutions whereas aragonite is deposited from hot solutions (warmer than 35 °C), which highly influences its internal structure. Its internal structure — which is different from that of calcite — accounts for its different crystalline form and physical properties.

Calcium carbonate — $CaCO_3$, orthorhombic.
Hardness: 3.5—4
Sp. gr.: 2.9—3.
Streak: colourless

1 — twins of aragonite crystals (Molina, Arago, Spain),
2 — sprudelstein (Karlovy Vary, Bohemia)

Aragonite

Large six-sided columns of aragonite are found in Aragon, Spain (hence its name). When, in the 15th century, a gallery was driven — into Erzberg near Eisenerz, northern Styria (Austria), in search of silver and gold veins — across a rich deposit of siderite ores, an interesting stalactitic coralloidal variety called Flos Ferri (flower of iron) was discovered, and is still being found there today. It consists of beautiful snow-white, divergent and ramifying branches. Outside Europe it may be found in Algeria.

Crystallized aragonite — much sought after by collectors — is of practically no economic importance. The sprudelstein, however, composed of brown and greyish bands of different shades, has been used since long ago for ornamental and decorative purposes, as has peastone. In Karlovy Vary, aragonite is deposited inside the water pipes draining mineral water from springs. Consequently, they must often be replaced by new ones. It gets deposited so quickly that in a few hours objects emerging in mineral water become coated with a crust of aragonite. In this way the so-called 'Carlsbad flowers' are formed, i.e. flowers encrusted with aragonite. The fine-fibrous *erzbergite* (mixture of aragonite and calcite) resembles the Carlsbad sprudelstein almost beyond distinction. It occurs in Erzberg, together with the Flos Ferri, and is also used as ornamental stone.

Calcium carbonate —
$CaCO_3$, orthorhombic.
Hardness: 3.5—4.
Sp. gr.: 2.9—3.
Streak: colourless

Flos Ferri
(Eisenerz, Styria)

134

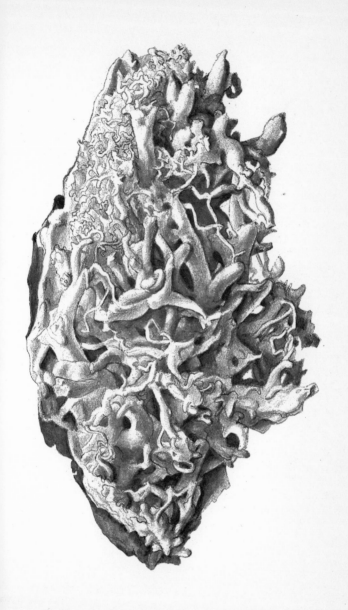

Magnesite

Magnesite got its name because of the content of magnesium. Like other carbonates, it occurs in nature in various forms, either crystalline or compact. The crystalline variety is fine- to coarse-grained, closely resembling marble. It is most often greyish with an admixture of dolomite; or yellowish, containing siderite. The compact variety is white. Granular magnesites are due to alterations of limestone and dolomite — less often, they originate by deposition; compact varieties are the result of decomposition of serpentines. They often contain admixtures of chalcedony or opal, being consequently harder. If they contain no admixtures, they are earthy, resembling chalk. The chief European deposits are in Austria (Veitsch, Leoben, and Mürzzuschlag, Styria), Czechoslovakia (between Lučenec and Košice, Slovakia), in the USSR (Satkinsk) and Greece, and outside Europe in the USA.

Magnesite is an important raw material used in the building industry. It is also widely made use of because of its fire-resisting qualities. It is used in the production of refractory bricks and furnace-linings. It is used in the production of porcelain and ceramic ware; large quantities are employed in the manufacture of the quick-setting cement, the so-called Sorel's cement.

Magnesium carbonate— $MgCO_3$, hexagonal.
Hardness: 4—4.5.
Sp. gr.: 2.96—3.12.
Streak: colourless to grey.

Magnesite
(Veitsch, Austria)

Dolomite; Ankerite

With its crystal shape, chemical composition and colour, dolomite resembles calcite and magnesite. Sometimes whole mountain ranges are composed of it. It may also include different admixtures, e.g. iron carbonate, and then it is called *ankerite*. Dolomite (ankerite less often) originated by deposition of shells of tiny sea animals. Their origin is therefore similar to that of limestone rocks. Crystallized dolomite and ankerite are common veinstones of metalliferous veins, where dolomite is the result of crystallization from hot solutions. Ankerite occurs in sediments of coal basins. Dolomite gave its name to a large mountain range, the Dolomites in northern Italy, which are predominantly composed of it. Crystallized varieties occur in Hall, Tirol, in the Binne valley, Switzerland, and in Vermont, the USA. Ankerite is found in Eisenerz, Styria and the Kladno and Ostrava coal basins, Czechoslovakia. Both minerals are used in the manufacture of special cements and for making refractory furnace linings. Ankerite, like calcite, is used as flux in furnace iron ore smelting.

Carbonates of calcium and magnesium: dolomite — $CaMg(CO_3)_2$, ankerite — $Ca(Fe,Mg)(CO_3)_2$, hexagonal. *Hardness:* dolomite 3.5—4; ankerite 3.5. *Sp. gr.:* dolomite 2.8—2.9; ankerite up to 3.3 (with maximum content of iron). *Streak:* dolomite colourless; ankerite grey.

1 — crystallized dolomite (Hall, Tirol),
2 — crystallized ankerite (Kladno area, Bohemia)

1

2

Talc

The name talc is derived from the Arabic; the name of its massive variety *steatite* (*soapstone*) is of Greek origin (*stear* — suet). It shows that this mineral of greasy or soapy feel was known to mineralogists long ago. Talc forms perfectly cleavable, flexible scales and plates of white or greenish colour and pearly subtransparent to translucent lustre. Its variety, steatite, often forms schistose rocks. Talc occurs in larger quantities in some deposits of crystalline carbonates, such as magnesites, in rocks rich in magnesium and in crystalline schists. The chief producer is the USA (in carbonate rocks in New York State, and in Vermont and Virginia in silicate rocks rich in magnesium). In Europe the largest deposits are in Italy (Piedmont) and Austria.

The so-called *potstone* (mixture of talc and chlorite) was used in the manufacture of vessels for cooking. Nicely coloured and harder varieties were carved into ornaments. At present, it is predominantly used as an important refractory material, e.g. in the manufacture of fire-resisting ceramic articles, especially tiles, vessels, and facing bricks for furnaces.

Hydrous magnesium silicate — $Mg_3(OH)_2(Si_2O_5)_2$, monoclinic — pseudohexagonal
Hardness: 1.
Sp. gr.: 2.6—2.8.
Streak: colourless.

Talc (Piedmont, Italy)

Gypsum; Anhydrite

Both are similar in form as well as chemical composition. Anhydrite, which got its name from the Greek *an* (without) and *hydor* (water), in contrast to gypsum, does not contain water. Gypsum occurs in columnar to tabular crystals resembling mica (Maria Glass). Pure crystallized gypsum is colourless, its massive variety being white or yellowish. Anhydrite is usually compact, rarely cleavable, whitish to bluish. Both minerals are formed by deposition from seawater. They usually occur associated, anhydrite sometimes altering to gypsum due to water absorption. The most important deposits of gypsum are near Volterra (Tuscany), in Spain and Egypt. Anhydrite is found in Slovakia and Poland.

Gypsum has been known since Ancient Times. The ancient sculptor Lysistratos was the first to use plaster in his work. Gypsum and anhydrite are mainly used in the production of plaster. Gypsum heated at 300—400 °C yields the plaster of Paris, which has the property of setting or becoming hard after being mixed with water. It is used in the building industry, medicine, and for casts. If heated to more than 400 °C, it does not absorb water. It is then used in paints and cement. The white, fine-grained variety — alabaster — is used for statues.

Calcium sulphates: gypsum — hydrated — $CaSO_4 . 2H_2O$; anhydrite — anhydrous — $CaSO_4$; gypsum monoclinic; anhydrite orthorhombic. *Hardness:* gypsum 1.5—2; anhydrite 3.0—3.5. *Sp. gr.:* gypsum 2.3; anhydrite 2.89—2.98. *Streak:* colourless to white, in anhydrite also grey.

1 — crystallized gypsum (Salzkammergut, Austria),
2 — fibrous gypsum (Volterra, Tuscany)

1

2

Quartz and its varieties

Quartz has always been a well-known mineral, much attention being paid to its purest variety — rock crystal. The name 'rock crystal', derived from 'crystal', originated from the Greek expression for ice. It got its name because of its close resemblance to ice, and because it keeps cool even in high temperatures. This is due to its internal structure which makes it a good conductor of heat. The wealthy ancient Romans knew of this property and kept spheres of rock crystal in their homes on which to cool their hands.

In nature, quartz appears in a great number of varieties, crystalline as well as in massive aggregates. Apart from the vitreous, colourless variety — the *rock crystal* — there are numerous kinds of quartz differing markedly in colour. The crystalline variety of varied shades of violet, *amethyst* (from the Greek *amethystos*, unintoxicating), used to be considered a remedy against intoxication. Other varieties are the crystalline yellow quartz called *citrine* (false topaz, Spanish topaz); the crystalline *smoky quartz* — a smoky-brown variety; and the black variety *morion* (an old Latin name, given originally by Pliny to black jasper and onyx from India). The rose-coloured variety is called *rose quartz*. Massive, granular as well as crystalline quartz of red, brown or yellow colour — caused by some iron oxides — is called *ferruginous quartz*. The so-called *milky quartz* is a pure milk-white, translucent or nearly opaque variety. The milk-white colour is caused by the refraction of light from numerous microscopic flaws. The grey

Silicon dioxide — SiO_2, hexagonal.
Hardness: 7.
Sp. gr.: 2.65.
Streak: colourless white to grey.

Crystals of rock crystal (St. Gotthard, Switzerland)

Quartz and its varieties

cat's-eye and yellow-brown *tiger's-eye* are quartz varieties which are pseudomorphic after the fibrous hornblende minerals. When cut in a round shape (en cabochon) they exhibit an chatoyant effect. Differently coloured, impure varieties are called *common quartz*. *Cap-quartz* is a crystalline variety of quartz containing scales of mica or other different admixtures distributed in a regular manner so that they may be separated into sections or layers. Under a hammer stroke, the crystals are easily divided into several 'caps'. *Jaspers* (old Latin name used by Pliny) are mixtures of chalcedony and quartz coloured by haematite to give red, yellow or green. Bright or dark green chalcedony with small spots of red jasper is called *heliotrope* (from the Greek *helios*, sun, and *trepa*, to change). It was used in ancient times, in observations of the Sun. A like mixture of chalcedony and jasper is the grey *chert* and the related *flint*, a darker to almost black variety of quartz.

Quartz is the commonest of minerals, forming 12 % of the Earth's crust. It is an essential constituent of many rocks, including sandstone. In nature it forms comparatively large crystals, especially in the cavities of coarse-grained granites, the so-called pegmatites and some ore veins, where it forms the gangue material. It often occurs in the form of isolated quartz veins. Crystals of rock crystal generally occur in cracks in crystalline schists, but also in igneous rocks, such as melaphyres, associated with agates. Amethyst may also be found in cavities in igneous rocks, especially in melaphyre

Silicon dioxide — SiO_2, hexagonal. *Hardness*: 7. *Sp. gr.*: 2.65. *Streak*: colourless, white to grey.

Druse of amethyst crystals (Rio Grande do Sul, Brazil)

Quartz and its varieties

amygdaloidal rocks in which it sometimes forms hollow spheres called geodes. It occurs as gangue in ore veins or as isolated quartz veins. Of a like occurrence is ferruginous quartz and morion. Rose quartz commonly occurs as a constituent of pegmatites. Some quartz varieties — often in association with chalcedony (e.g. jasper) — originated in cavities in volcanic rocks, especially melaphyres. The constituents of volcanic rocks were decomposed under the effects of hot solutions and the cavities were filled with silica solution. Chert originates in ore veins as a mineral of unorganic origin; flint has been formed of silicate shells of animals extinct in the Mesozoic Era (Jurassic and Cretaceous Periods).

The most beautiful crystals of quartz, especially rock crystal, were found in crystalline schists in the Alps. In 1735, the Rock Crystal Vault was discovered in Zinggenstock in the vicinity of Grimsel, Bernese Alps, Switzerland. It is a large cavity which has yielded more than 1000 q of rock crystal. The largest amethyst geode was found in Brazil, measuring $10 \times 5 \times 3$ metres. Similar, although much smaller, cavities, filled with crystalline as well as amorphous varieties of quartz, may be found in Idar, Federal Republic of Germany, and in the area of the Giant Mountains, near Kozákov and Levín, Bohemia. Large amounts of flint may be found along the coast of the Baltic Sea, especially on Rügen Island.

Silicon dioxide — SiO_2, hexagonal.
Hardness: 7.
Sp. gr.: 2.65.
Streak: colourless, white to grey.

Crystal of smoky quartz
(St. Gotthard, Switzerland)

148

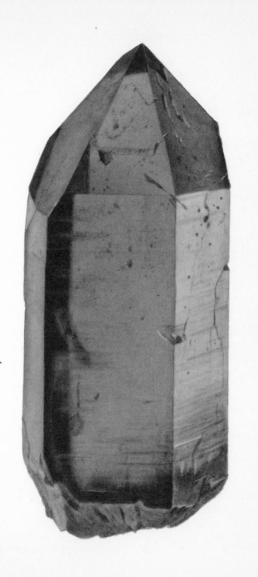

Quartz and its varieties

On account of its hardness, power of resistance and a comparative cheapness, rock crystal is used for larger ornamental works. Also amethyst has been a favourite precious stone and raw material for the manufacture of amulets, being used in jewellery and for making ornamental objects since Ancient Times. Its bright, evenly coloured crystals are extremely valued, and when heated change into honey-yellow artificial citrines, called also *Spanish topaz*. They are considered to be precious stones. At present, quartz is an important raw material for the glass and optical industry. The transparent variety of rock crystal is used in the manufacture of optical glass. It is also of importance in radio technology. For the latter purpose, however, artificial (synthetic) crystals of quartz are applied.

Silicon dioxide — SiO_2, hexagonal.
Hardness: 7.
Sp. gr.: 2.65.
Streak: colourless, white to grey.

1 — rose quartz (Zwiesel, Bavaria),
2 — Tiger's Eye Doorn Mount., South Africa)

Chalcedony:
Carnelian, Cacholong, Plasma, Agate

These are all sub-varieties of silicon dioxide which are— contrary to quartz— microscopically crystalline, i.e. their crystals are so small that they can be seen only under a strong microscope if cut into very thin translucent sections. Chalcedony always appears compact, occurring in the form of reniform crusts, mammillary and bedded coatings, and other forms. Its colour varies greatly. *Common chalcedony* forms grey, bluish or yellowish translucent aggregates. *Carnelian* is coloured red by iron oxides; *cacholong* is a chalk-white mixture of chalcedony and opal; *plasma* is a light green variety. The best known and most popular variety of chalcedony, however, remains the finely banded *agate*. It is most often found in amygdaloidal cavities in the melaphyres and similar rocks. These rocks were formed by volcanic activity culminating in eruptions of molten vesiculose, gas-rich lava. Compact chalcedony fillings, as well as concentrically arranged agates, were formed then by deposition from hot solutions in numerous vesicles left behind after the escape of gases. These fillings, which are much harder and more resistant than the rock itself, may also be found in arable soil formed by the decomposition of the original rock.

Silicon dioxide — SiO_2, hexagonal, cryptocrystalline.
Hardness: 6.5.
Sp. gr.: 2.57—2.6 (lower than that of quartz).
Streak: colourless, white to grey.

1 — stalactitic chalcedony (Faroe Islands), 2 — carnelian (Nová Paka, Bohemia)

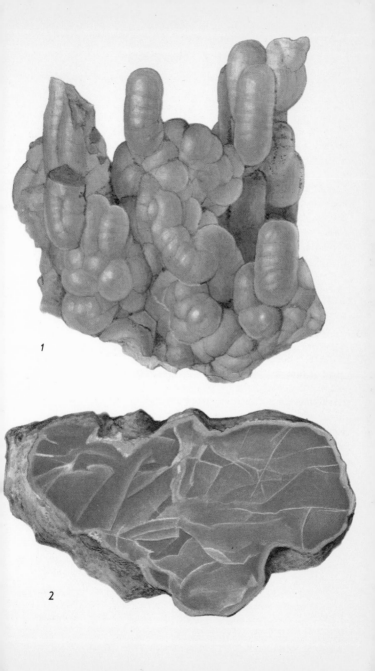

1

2

Chalcedony

Agate was named after the first known locality along the river Achates (today Dirillo) in southern Sicily. The present most important deposits are in Brazil and Uruguay; in Europe in Idar, Federal Republic of Germany, and in the Giant Mountains, Bohemia, where chalcedony was found as early as the reign of the Emperor Charles IV. It was used for ornamental purposes. During the reign of the Emperor Rudolf II, the interest in chalcedony grew to such an extent that the Emperor employed his own collectors from remote foreign countries and provided them with special letters of safe conduct and other privileges.

Agate, one of the oldest known minerals, was used for ornamental purposes as early as 5000 B.C. Although agate and chalcedony jewels are not as popular today as they used to be in the past, both these minerals are used in the manufacture of different articles of artistic merit, e.g. vessels, cups, boxes, writing sets, decorative paper-weights, ash-trays, etc. At present, both agate and chalcedony are also used in various industrial branches. On account of their hardness and power of resistance, massive pieces of these minerals serve as excellent raw materials in the manufacture of bearings and blades in laboratory scales, compasses, and other instruments; mortars and spherical bearings for dairy machines.

Silicon dioxide — SiO_2, hexagonal, cryptocrystalline.
Hardness: 6.5.
Sp. gr.: 2.57—2.64 (lower than that of quartz).
Streak: colourless, white to grey.

Agate
(Rio Grande
do Sul, Brazil)

Diamond

This is one of the modifications of carbon and has thus the same composition as, for instance, graphite. The enormous difference in the properties of these two minerals is due to the different atomic arrangement. In diamond the atoms of carbon are arranged in layers very close to each other. This is held to account for its physical properties, such as the highest known mineral hardness. It usually forms rounded octahedra, often twinned, transparent, yellow, brown, green, less often blue and black; it rarely occurs in the form of hexaoctahedra. Its crystals have adamantine lustre. Diamond occurs in sediments of ultrabasic rocks poor in silicon dioxide. The only known rock in which diamond has been found ingrown is kimberlite, named after the locality of Kimberley, South Africa. Other important deposits are in Zaire, Angola, Sierra Leone, Brazil, and the USSR (Yakutia).

Diamond, which is the most valued precious stone, is used for gems (cut to the form of brilliant), as well as for technical purposes. Diamonds of inferior quality are called *bort*. They are used as abrasives, in rock boring (diamond bits), and as dies for wire drawing.

Carbon — C, cubic.
Hardness: 10.
Sp. gr.: 3.5—3.6.
Streak: colourless

Diamond crystal in kimberlite (Kimberley, South Africa)

Corundum:
Ruby, Sapphire, Leucosapphire

Corundum has several varieties which are due to the admixtures of different metals. The red *ruby* is coloured by chromium, the blue *sapphire* by iron and titanium. The colourless, perfectly clear variety is called *leucosapphire*. Corundum may occur in yellow, green and violet colours. *Common corundum* is usually grey. It occurs in crystal as well as granular form, disseminated in igneous rocks or crystalline limestones, being frequent also in alluvial deposits. The chief producers are South Africa, India, Madagascar, and Canada; in Europe the largest deposits are at Mijas in the Urals, at Unkel in the Rhineland, and at Jizerská louka, Bohemia, from which come the most beautiful sapphires in Europe.

Corundum, or more precisely, some of its nicely coloured varieties, are precious stones. On account of its hardness (9) it is used as an abrasive, and for some other technical purposes. It facilitated the construction of lasers, which amplify light by stimulated emission of radiation over long distances. The main component of the laser — which is of extreme importance for peaceful as well as military purposes — is the ruby. For these purposes, however, synthetic rubies may also be used.

Aluminium oxide — Al_2O_3, hexagonal.
Hardness: 9.
Sp. gr.: 4.1.
Streak: colourless to grey.

1 — crystals of ruby (Madagascar), 2 — crystals of sapphire (Montana, USA)

1

2

Beryl:
Emerald, Aquamarine, Golden Beryl, Morganite

According to the Roman naturalist Pliny, the Emperor Nero used emerald lenses as spectacles. This variety was most probably cited here by mistake and for that purpose some other variety was used. The *emerald* is green; *aquamarine* is pale blue, having the colour of sea water. Varieties of other colour shades are less frequent, such as the yellow, transparent variety called *golden beryl*, and the pink variety called *morganite. Common beryl* is of an inferior quality, unfit for gem purposes. It forms crystals and columnar aggregates and occurs as an accessory mineral in coarse-grained granites called pegmatites, and other similar rocks. The best emeralds occur chiefly in calcite veins and mica-schists in Colombia, the Urals and Peru, and in Europe in the Salzkammergut. Aquamarines are found in Brazil, the common beryl in Maine (USA).

Since Ancient Times, beryl has been considered a highly-prized gemstone. Emeralds of perfect colour are sometimes more valued than diamonds. Also, other varieties of transparent beryls are used as gemstones. Apart from being valued as a gemstone, it is an important raw material of beryllium, which is especially important in the production of light alloys.

Beryllium aluminium silicate — $Al_2Be_3Si_6O_{18}$, hexagonal.
Hardness: 8.
Sp. gr.: 2.6—2.8.
Streak: colourless, white to grey.

1 — crystals of emerald (Musso, Colombia),
2 — crystals of aquamarine (Minas Gerais, Brazil)

Tourmaline:
Schorl, Rubellite, Indicolite, Verdelite, Dravite, Achroite

In 1703, Dutch sailors brought to Europe a special stone from Ceylon called *turamali* by the native inhabitants. The stone exhibited a strange property: heated specimens attracted the ash from the fire on which they had been heated. In tourmaline electricity originates, on account of its special internal structure. The development of plus (+) and minus (−) electricity at the opposite ends of the tourmaline columns corresponds to the dissimilar molecular structure. The Ceylon variety is reddish to slightly red *(rubellite)*; the less conspicuous black variety *(schorl)*, however, had been known a long time before. There is also an indigo-blue variety *(indicolite)*, a green variety *(verdelite — Brazilian emerald)*, a brown variety *(dravite)*, a colourless *(achroite)*, and many coloured varieties: the so-called 'Moor's or Turkish heads'. It occurs in long-, less often short-columnar crystals, vertically striated, of approximately three-sided cross section. More frequently it is found in radiating needles, especially in granites and pegmatites. In Ceylon it is found in sedimentary rocks. Other localities are Pala in San Diego County, California; Brazil, and Madagascar. The chief European localities are in the Urals (USSR). Stones of good colour are used as gems. It is also used in optics (dark, transparent varieties).

Complex borosilicate of a whole series of metals, especially aluminium, sodium, magnesium and lithium, hexagonal. *Hardness:* 7 and more. *Sp. gr.:* 3—3.2. *Streak:* colourless to grey.

1 — crystal of tourmaline — schorl (Dolní Bory, Moravia) 2 — crystals of rubellite (Pala, California)

Topaz

As a gemstone it has been popular since Ancient Times. In nature it most often occurs colourless, white or grey. Some metalliferous admixtures cause its different colouring, i.e. it may be found in yellow, blue, pink or other colours. It occurs in the form of columnar crystals. More often, however, it is granular or columnar (so-called pycnite). It is a characteristic mineral of some granites, especially their coarse-grained varieties (pegmatites). It is generally associated with cassiterite. Excellent crystals are found in pegmatites in the Urals (Murzinka) and in Brazil (Villarica).

Topaz is a precious stone which has remained popular since the earliest times. This is especially due to its hardness, transparency, comparatively large crystals, flawless surface, and beautiful colour shades. Some varieties, especially richly coloured ones, fade on exposure to sunlight. Since it has been shown that it is not the light but the warmth of the sun's rays which causes the change of colour, jewellers started changing its colour artificially by heating it.

Aluminium fluosilicate — $Al_2F_2SiO_4$, orthorhombic. *Hardness*: 8. *Sp. gr.*: 3.5. *Streak*: colourless to grey.

Crystal of topaz (Minas Gerais, Brazil)

Garnets:
Pyrope and Almandite

The garnet group embraces minerals of similar composition and structure, grading over into one another. The best known garnets are *pyrope (Bohemian garnet)* and *almandite (Oriental garnet)*. Pyrope usually occurs in grains of beautiful deep red colour and vitreous lustre. It originates in basic igneous rocks rich in olivine, especially kimberlites, and serpentines which are due to alteration of the latter. Important localities are the Central Bohemian Highlands, where it is found in alluvial deposits, and South Africa. Almandite is more frequent than pyrope and is found in larger crystals, bounded by planes of the rhombododecahedron. They are usually pale violet. It most often occurs in gneiss and granulites. It is named after its original locality in the vicinity of Alabanda, Asia Minor, where it used to be extracted in Ancient Times. Chief European localities are in the Tirol.

Garnets are popular gemstones used as decorative jewels. For gem purposes especially, Bohemian garnets are required. Almandite is the more valuable, since it closely resembles ruby.

Duplex silicates trivalent and divalent metals. Pyrope — magnesium-aluminium garnet — $Mg_3Al_2(SiO_4)_3$; almandite — iron-aluminium garnet — $Fe_3Al_2(SiO_4)_3$, cubic.
Hardness: 7 and more.
Sp. gr.: pyrope 3.7—3.8; almandite 4.1—4.3.
Streak: colourless

1 — almandite (Tirol),
2 — pyrope (Central Bohemian Highlands)

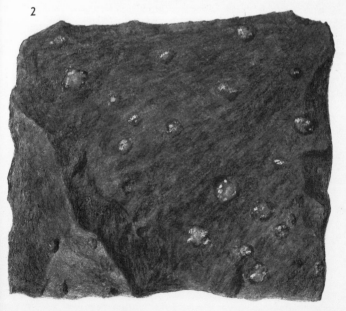

Opal:
Common Opal, Precious Opal, Fire Opal, Dendritic Opal, Milk Opal, Hydrophane, Hyalite, Menilite (Liver Opal), Cacholong

Opal is one of the minerals which exhibit opalescence and an excellent play of colours. Due to its brilliant colour it was one of the first precious stones used for ornamental purposes, and as early as Ancient Times it was highly valued. The oldest preserved objects made of opal date from approximately 500 B.C. The Roman senator Nonius, who possessed a finely engraved opal, was exiled after he had refused to pass it on to the Emperor Marcus Aurelius.

The play of colours, which is characteristic of the precious varieties of opal, is due to the refraction of light from finely dispersed cracks filled with water. Varieties displaying an excellent play of colours are called *precious opals*. Some opals lose part of their water content in dry air and consequently also the play of colours. Immersed in water or in damp air they may retain their content of moisture. They are called *hydrophanes*. Yellow-red opal displaying a beautiful play of colours is called *fire opal*. Opaque varieties are called *common opals*. They are compact and show many colours. If interwoven with black dendrites of manganese and iron oxides, they are called *dendritic opals*. *Hyalite* is a transparent colourless glassy variety (from the Greek *hyalos*, glass) occurring in reniform, botryoidal or stalactitic forms. *Menilite* (called after its source, Ménilmontant on the slopes of Montmartre, Paris) is a grey-brown liver-coloured variety found in flattened or rounded concretions similar to flint.

Silicon dioxide with a varying amount of water $SiO_2 + nH_2O$, amorphous. *Hardness:* 5.5—6.5. *Sp. gr.:* 1.9—2.3. *Streak:* colourless to grey.

1 — common opal (southern Slovakia), 2 — fire opal (Zimapán, Mexico)

168

Opal

Cacholong is a chalk-white variety mixed with chalcedony. Opal originates either by deposition from hot solutions, such as hot springs, or by decomposition of silicates in rocks, especially serpentines. In a similar manner it also originates in andesites and basalts. Opal rocks named *tripolites (diatomaceous earth, diatomite)* after their localities in Tripoli, North Africa, were formed by deposition of siliceous organisms such as the algae and diatoms (hence their name). Menilite and cacholong are of predominantly organic origin. Up to the seventies of the last century the most beautiful opals were those found at Dubnik near Prešov, Slovakia. In this locality of ancient origin the best specimens of precious opals were found. The fame of Slovakian opals was overshadowed by the discovery of opals in Australia and Mexico. In Mexico, fire opals are also found. Diatomaceous earth is used as a polishing powder for metals, in the manufacture of dynamite, as a heat insulator and noise killer, and in the manufacture of filtering and adsorption material.

Silicon dioxide with a varying amount of water $SiO_2 + nH_2O$, amorphous.
Hardness: 5.5—6.5.
Sp. gr.: 1.9—2.3.
Streak: colourless to grey.

1 — hyalite (Valeč, Bohemia)
2 — menilite (Paris — Ménilmontant, France)

1

2

Hornblende, Amphibole:
Common Hornblende,
Basaltic Hornblende

It is a rock-forming mineral closely resembling augite. Its chemical composition, crystal form, and colour are strikingly similar to those of augite. With a little experience, however, they may easily be distinguished. They are both of comparatively good prismatic cleavage, giving two sets of cleavage-planes meeting at an angle of nearly 90° in augite, and 120° in hornblende. Crystals of hornblende usually are short- to long-columnar, terminated by three faces. They are bounded by almost perfectly smooth and straight plane faces. It also forms irregular aggregates of granular, columnar or fibrous crystals. Its colour changes according to its variable chemical composition. The so-called *basaltic hornblende* is black, in thin sections brown. *Common hornblende* is black to black-green. It is abundant in nature and occurs in many rocks poor in silicon dioxide. Most often it can be found in igneous rocks, especially of the Tertiary age, predominantly in tuffs and basalts. Brilliant crystals are found on the hill Vlčák near Černošín in the vicinity of Stříbro, Bohemia, and in the Central Bohemian Highlands, especially in the vicinity of Lovosice.

Very complex silicate of aluminium, magnesium, sodium and calcium. Basaltic and common hornblende are monoclinic; some sorts of hornblende crystallize in the triclinic system. *Hardness:* 5—6, in magnesium-rich specimens up to 6.5.
Sp. gr.: 2.9—3.6.
Streak: green-grey.

1 — crystals of hornblende (Vlčák, Bohemia), 2 — crystal of hornblende

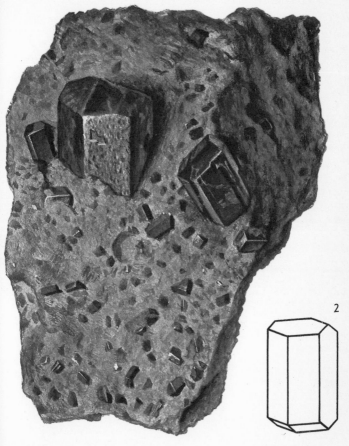

1

2

Augite

Augite belongs to those rock-forming minerals that so far have no practical application, but are quite abundant in nature, such as quartz, feldspars, micas and hornblende. It occurs in short-columnar or tabular crystals, as well as disseminated grains of black colour with brownish or leek-green tint. It sometimes occurs as contact twins. It is strikingly similar to hornblende in its chemical composition, crystal form, and colour. With a little experience, however, they may easily be distinguished. Both these minerals have a good prismatic cleavage, giving two sets of cleavage-planes meeting at an angle of nearly 90° in augite, and 120° in hornblende. It is found in many rocks poor in silicon dioxide, especially basaltic tuffs and other basaltic rocks, in andesites and many volcanic rocks, especially of the Tertiary age. Individual crystals of augite are often found in arable soil in the vicinity of parent rocks. Collectors especially appreciate crystals from basaltic tuffs from the Central Bohemian Highlands, Bořislav near Teplice, Černošín in the vicinity of Stříbro, Bohemia; and from Kaiserstuhl, Germany.

Very complex silicate of aluminium, magnesium, sodium and calcium with admixture of titanium, monoclinic. *Hardness:* 5—6. *Sp. gr.:* 3.2—3.6. *Streak:* greyish-green.

1 — crystals of augite (Paškopole, Central Bohemian Highlands), 2 — crystal of augite

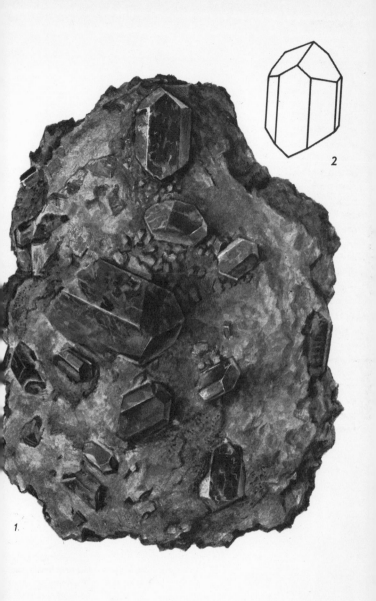

2

1.

COLLECTING MINERALS

For every collector, the initial stage in collecting minerals begins in the home, as every collecting trip must be carefully planned beforehand. In the first place, it is necessary to study some special mineralogical and geographical literature to gain a little knowledge of the character of the chosen locality. A preliminary study of the locality will be impossible without a good map, especially a good geological map. Apart from the theoretical preparation, the practical part is of no less importance. It is necessary to have a good ruck-sack, good geological hammers — one large and one small — a chisel, maps, a notebook, a hand-lens, and some paper for wrapping up the collected minerals. We must not forget to take small labels which will serve the purpose of marking the locality where the specimen was found. It is also useful to take a camera.

An indispensable aid in field work as well as at home is a good hand-lens which is often very important in the correct identification of minerals. Collectors should use a mineralogical

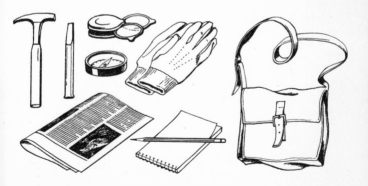

Fig. 9. The most important equipment for mineral collectors: geological hammer, chisel, lens, compass, newsprint for wrapping up mineral specimens, notebook with pencil (not ink pen!). bag.

177

lens — magnifying 8- to 10-times — which eliminates any distortion of the visual field. In this way it differs from other kinds of commonly applied lenses, e.g. botanical ones.

It might seem surprising that the most suitable paper for wrapping up the specimens is newsprint. There is no need for any special paper bags or soft paper. For wrapping up finer materials, it is advisable to have some small paper trays of different sizes. Their many-sided application will be described later.

Mineral localities may occur anywhere where the Earth's crust becomes exposed in outcrops, such as in quarries, mines, and coal piles. Good localities may also occur in places where roads and railway tracks cut down into the relief, especially in places of fresh exposure. Minerals may also be found along roads, in arable soil, in alluvial deposits and elsewhere. In all cases the collector should ascertain that the locality is the original source of the mineral and not a place to which it was brought by Man.

The majority of collectors collect only mineral specimens, although collections of rocks are also of great interest. It is a pity that collections of rock samples are comparatively rare. Let us say a few words about work in the field, the collecting of mineral specimens, and finally let us come back to the less popular rocks.

The collector finds it easier to orientate himself on coming to a locality if he is already familiar with it from the literature studied at home. It is, however, always advisable to make notes regarding the locality, the state of the rocks and their occurrence, as well as the mineral specimens to be found there. Naturally the date of finding should be added. It is also useful to make a small sketch of the locality complete with drawings or documentary photographs — showing a general view, as well as details.

Such notes may be of great importance, especially later, when the collector wants to rearrange and reclassify the collected specimens at home, or in case he wishes to visit the locality again. Therefore notes should be made in a manner which allows easy reference to the precise locality of individual minerals. In this way the notes become a documentation of the locality in question, at a specific time, and facilitate the study of its contingent changes.

It is better to collect specimens of a certain size, the most useful size being 6 × 9 cm. Of course, if we found a larger crystal or a large group of crystals, we should never break it up to get the required size. Also, smaller loose crystals, or a smaller but characteristic specimen should not be discarded.

The choice of minerals should be made in situ, i.e. the collector should try to find a typical specimen of every mineral. If a mineral occurs in the locality in several different forms, such as crystalline, massive, in various aggregates and differently coloured varieties, the best specimens of every type should be chosen. From the scientific point of view, those specimens that are obviously related to the parent rock, or to other associated minerals, are of special importance. Such specimens characterize best the mineral assemblage of the locality in question (paragenesis). The trimming of specimens should be carried out directly on the spot, since there is always the hope of finding another specimen if the original gets destroyed in this process. In trimming a mineral specimen, its development, typical forms and the above-mentioned genetic relationship should be observed. At the same time, of course, the collector tries to achieve the best possible appearance, so that the specimen will look well in his collection.

All the specimens should immediately be carefully labelled, particularly as regards the locality. If a specimen is found in a stone quarry, the notes should comprise the description of the particular spot; if it is found in an exposure on a road or railway track, the estimate of the distance to the nearest village, or some other pertinent data should be added, according to the map. Such topographic data are very important since a well-arranged and smart collection includes only specimens provided with exact description of the localities in which they were found.

Every specimen, once provided with a label, should be carefully wrapped up in newsprint and placed into the ruck-sack. It does not pay to try to spare the wrapping paper since insufficiently wrapped specimens may easily be damaged during transit. Much attention should be paid to fine crystalline or hair-like forms of crystals. They should be placed in small paper trays or cardboard boxes where they must be fastened by means of a thread. Using cotton wool for wrapping up the specimens is not always advantageous since cotton may sometimes damage the finest crystals.

A few words must be said about collecting rocks, to which amateur collectors pay but little attention. This is probably because of the less conspicuous forms and colours of rocks, and perhaps, for the layman, the comparative difficulty of identification. Rock specimens are most useful in completing the picture of the locality. The occurrence of minerals in nature is never accidental but lawful, and minerals always have a certain correlation to the rocks in which they occur.

Rock samples are collected in the same manner as mineral specimens. Only their size — as far as possible — must be observed more carefully. Trimming of samples to the required form and size is carried out by means of a geological hammer. The most useful size is 6×9 or 9×12 cm, thickness 3—4 cm. A smaller fragment of the rock is always added, to be at hand in case of different tests (e.g. sections). Loose rocks are placed directly into boxes of corresponding size or into glass tubes to fit the size of the box.

In conclusion, a few words must be said about handling the collected mineral specimens and rock samples at home, as well as the arrangement of collections.

The greatest problem is the correct identification of minerals. Some minerals cannot be identified by ordinary simple means without the aid of some measuring apparatus. The identification of minerals requires some theoretical knowledge and experience. Usually the amateur collector is able to identify the most common mineral species from his locality on the basis of their description, after a preliminary study of some literature.

At the same time attention must be paid to the physical properties of minerals, which are either obvious at first sight (mineral modification and colour), or are easily determined by some simple means (streak and hardness). Minerals are of metallic, submetallic or unmetallic lustre. The modification of submetallic minerals is usually not constant, they may sometimes be metallic, sometimes unmetallic. When identifying a mineral according to its colour, the collector should bear in mind that one and the same mineral may occur in various colours at different times. Moreover, every mineral must be expected to occur in a colour differing from that described in the textbooks.

This book also includes advice on how to determine streak and hardness. In the pictorial part, hardness of individual minerals is given. Because of its limited extent, the book should not, however, be used as a mineralogical key. In the identification of minerals, transparency, kind and intensity of lustre, cleavage and fracture of minerals are observed. If the collector has some heavy liquid of a given specific gravity at his disposal, he may, by simply submerging the mineral specimen, determine the latter's approximate specific gravity.

In more complicated cases, of course, it is advisable to consult a more experienced collector, or to seek advice of specialists in different mineralogical institutes and museums.

The best way of keeping minerals is to preserve them in a cabinet. Mineral specimens and rock samples should never be placed loose in the drawers but in paper trays. To prevent the moving of trays each time the drawer is opened, it is useful to have trays of a definite size. The most useful size is 6×9 cm. Every collection includes larger, as well as smaller, specimens and therefore it is necessary to possess trays of different sizes. To fill the space of the drawer completely, the size of the larger trays should be a multiple of the basic size, i.e. 9×12 cm, and 12×18 cm, and of the smaller 4.5×6 cm. The most useful height of the trays is 3 cm and 1.5 cm respectively.

Every mineral is provided with a label which replaces the one written in situ. It should bear the name of the mineral, the exact source, the date of collecting and the manner of acquisition. Provisional labels may be left under the final labels. Many collectors also add some further data, such as a brief description of the specimen, its chemical formula or crystal system.

The arrangement of the collection is of extreme importance for every collector. There are two possible ways of arranging minerals, i.e. according to the mineral system or according to the locality. For a local collector the second may be better, but for a collector wishing to possess representative specimens of the whole mineralogical system without considering the locality, the first may be more advantageous. In both cases, however, a list of all minerals or rocks included in the collection is useful.

Table of Geological Ages and Systems

Age		System	Duration (in millions years)	Age
Quaternary			1	
				1
Tertiary	Late (Upper) (Neogene)	Pliocene	10	
				11
		Miocene	14	
				25
	Early (Lower) (Paleogene)	Oligocene	15	
				40
		Eocene	20	
				60
		Paleocene	10	
				70
Mesozoic		Cretaceous	70	
				140
		Jurassic	45	
				185
		Triassic	40	
				225
Paleozoic	Later (Upper) (Permo-Carboniferous)	Permian	45	
				270
		Carboniferous	70	
				340
	Early (Lower)	Devonian	50	
				390
		Silurian	40	
				430
		Ordovician	50	
				480
		Cambrian	90	
				570
Pre-Cambrian			1230	
				1800
Eozoic			1700	
				3500

GLOSSARY

Amorphous minerals — minerals devoid of lawful internal structure and consequently of crystallinity.

Amygdaloidal cavities — cavities formed in some igneous rocks by gases escaping from the lava.

Cleavage — property of many crystalline substances of breaking or splitting in definite directions.

Cleavage fragments — bodies bounded on all sides by cleavage planes.

Clerici's Solution — binary salt of thallium formate and malonate. Deadly poison!

Concretions — rounded masses formed secondarily in rocks by accumulation about a centre.

Contact minerals — minerals originating at contacts of igneous and sedimentary rocks.

Crystal — a body bounded by surfaces, usually flat, arranged on a definite plan which is an expression of the internal arrangement of ions, atoms, and molecules.

Crystal aggregate — group of crystals of one mineral of irregular growth.

Crystal system — all the crystal forms that can be referred to the same set of crystallographic axes.

Crystalline rocks — general term denoting igneous and metamorphic rocks.

Crystalline system — formation composed of crystalline rocks.

Crystallography — the study of crystals.

Dendrites — dark fern-like, branching coatings in fissures, composed of iron and manganese oxides.

Density of minerals — number denoting how many times the substance of a mineral is heavier than an equal volume of water.

Double refraction — the separation of a ray of light passing through certain crystalline substances into two rays.

Druse — closely crowded crystals of approximately parallel growth, growing from a common base.

Geode — cavity lined or completely filled with minerals, often well crystallized, usually of round form.

Geology — science of material composition and structure of the Earth's crust and its development.

Goniometer (cryst.) — apparatus for measuring interfacial angles of crystals.

Gossan — upper part of deposit formed by decomposition of original mineral particles due to atmospheric influences.

Guano minerals — minerals originating from layers of guano, i.e. fossil excrement of sea fowl.

Hardness — the degree of resistance which a mineral offers to another mineral pending to scratch it.

Intaglio — gem with incised design.

Interference of light — superposition of light rays of different wave lengths.

Isomorphism — replacement of related elements in compounds crystallizing in similar forms.

Karst phenomena — forms due to dissolution of limestones and mechanical activity of water in limestones.

Luminescence — glowing of some minerals due to exposure to light rays, rubbing or heat.

Magma — liquid molten siliceous rock substance. Its continuous cooling results in the origin of igneous rocks.

Massive minerals — minerals formed of comparatively large pieces of a single crystal.

Metallic colours — colours of opaque minerals of metallic lustre.

Metasomatism — replacement of the original mineral by another, usually less soluble.

Mineral — a substance having a definite chemical composition and atomic structure and formed by inorganic processes.

Mineral modification — general optical properties of minerals, especially colour and lustre.

Mineralogy — the study of minerals.

Mohs's Scale of Hardness — set of ten minerals used as standard examples for hardness tests. It's author is F. Mohs.

Nodules — general term denoting all nodular forms in sedimentary rocks.

Oolites — small, predominantly rounded bodies produced by the deposition of calcium carbonate in successive layers around small nuclei.

Palaeontology — science of the life in the past geological time as studied through the fossils.

Paragenesis — occurrence of associated, genetically related minerals which have originated gradually or simultaneously in one natural process.

Pedology — study of soils.

Penetration twin — twins of interpenetrated crystals.

Petrography — study of rocks.

Pleochroism — property of some minerals which display different colours when viewed from different directions by transmitted light.

Polymetallic deposits — deposits where silver, lead, zinc, and copper occur in association.

Propylitization — hydrothermal alteration by heated solutions of dark minerals and feldspars of some igneous rocks into secondary silicates.

Pseudomorphs — crystal forms of minerals in which original particles were substituted by new particles.

Radioactive irradiation — irradiation due to disintegration of atoms.

Recent minerals — minerals formed at present predominantly by deposition.

Rocks — individual portions of the Earth's crust.

Rohrbach's Solution — solution of mercury-barium iodide. Deadly poison!

Scalenohedron — crystal form of 8 or 12 faces, each face being scalene triangle.

Sclerometer — instrument for measuring hardness.

Skarn — metamorphosed siliceous rock rich in calcium and iron.

Skarn minerals — the main components of skarn, especially garnets, pyroxenes and magnetites.

Streak — the colour of the fine powder usually obtained by rubbing the mineral on unglazed porcelain.

Tarnish colours — iridescent mixture of colours of metallic minerals due to the interference of rays of light reflected from the weathered surface.

Thoulet's Liquid — solution of mercury iodide and potassium iodide in water. Deadly poison!

Twin — lawful, symmetrical intergrowth of two crystals.

Zeolites — hydrated silicates of univalent and divalent elements. Their internal structure enables water to penetrate the crystal or to be given off without disrupting the crystal structure.

BIBLIOGRAPHY

Bauer, J. and A.: *A Book of Jewels*. Artia, Prague, 1966.

Borner, R.: *Minerals, Rocks and Gemstones*. Oliver and Boyd. Edinburgh, 1966.

Bottley, E. P.: *Rocks and Minerals*. Octopus Books, London, 1972.

Desautels, P. E.: *Mineral Kingdom*. Hamlyn, London, 1971.

Evans, E. K.: *Rocks and Rock Collecting*. Golden Press, Western Publishing, Wisconsin, 1970.

Heavilin, J.: *Rocks and Gems*. Macmillan, New York, 1964.

Irving, R.: *Rocks and Minerals*. Knopf, New York, 1956.

Pough, P. H.: *A Field Guide to Rocks and Minerals*. Houghton Mifflin, Boston, 1963.

Simpson, B.: *Rocks and Minerals*. Pergamon, New York, 1966.

Simpson, B.: *Minerals and Rocks*. Octopus, London, 1975.

Talent, J.: *Minerals, Rocks and Gems*. Tri-Ocean, San Francisco, 1970.

Zim, H. S. and Shaffer, P. R., *Rocks and Minerals*. Hamlyn, London, 1966.

INDEX